Technische Universität München

Fakultät für Elektrotechnik und Informationstechnik

Lehrstuhl für Elektrische Antriebssysteme
und Leistungselektronik (EAL)

Noise Optimized Control of an Electrical Drive with Induction Machine

Harmonics Modeling and Control Strategies

Wolfgang Michael Bischof

Vollständiger Abdruck der von der Fakultät für Elektrotechnik und Informationstechnik der Technischen Universität München zur Erlangung des akademischen Grades eines

Doktor-Ingenieurs

genehmigte Dissertation.

Vorsitzender: Prof. Dr.-Ing. Bernhard U. Seeber

Prüfende der Dissertation:

1. Prof. Dr.-Ing. Dr. h. c. Ralph Kennel
2. Prof. Dr. ir. Dr. h. c. Rik W. De Doncker (RWTH Aachen)

Die Dissertation wurde am 07.11.2018 bei der Technischen Universität München eingereicht und wurde durch die Fakultät für Elektrotechnik und Informationstechnik am 07.06.2019 angenommen.

Noise Optimized Control of an Electrical Drive with Induction Machine

Harmonics Modeling and Control Strategies

Wolfgang Michael Bischof

Bibliografische Information der Deutschen Nationalbibliothek
Die Deutsche Nationalbibliothek verzeichnet diese Publikation in der Deutschen
Nationalbibliographie; detaillierte bibliographische Daten sind im Internet über
http://dnb.d-nb.de abrufbar.
1. Aufl. - Göttingen: Cuvillier, 2019
    Zugl.: (TU) München, Univ., Diss., 2019

© CUVILLIER VERLAG, Göttingen 2019
    Nonnenstieg 8, 37075 Göttingen
    Telefon: 0551-54724-0
    Telefax: 0551-54724-21
    www.cuvillier.de

1. Auflage 2019
Gedruckt auf umweltfreundlichem, säurefreiem Papier aus nachhaltiger Forstwirtschaft.

    ISBN 978-3-7369-7066-3
    eISBN 978-3-7369-6066-4

Danksagung

Zu dem Themengebiet der Modellierung und Regelung des akustischen Verhaltens von elektrischen Antriebssystemen fand ich durch meine Tätigkeit als Masterand bei der Firma Robert Bosch GmbH. Hierbei betreute mich Martin Hennen, welcher zugleich für diese Dissertation den Weg bahnte und diese anschließend betreut und jeder Zeit unterstütze. Martin Hennen gilt hierfür meine vollste Dankbarkeit. Ebenso gilt mein vollster Dank Matthias Bösing, welcher zu jeder Zeit ein offenes Ohr zum Arbeitsstand hatte, sowie meine Arbeit mit Rat und Tat unterstütze. Da diese Promotion durch die Firma Robert Bosch GmbH finanziert wurde gilt meine Erkenntlichkeit ebenso meinen damaligen und heutigen Vorgesetzten, welche offen für diese Themenstellung waren und diese unterstützen. In der Zeit der Erstellung dieser Arbeit bereute ich Bhaskar Chatterjee und Armin Sarcevic in ihrer Tätigkeit als Masteranden. Auf Basis ihrer Ergebnisse konnten ich die Qualität und Effizient dieser Arbeit deutlich steigern, weshalb ich Ihnen ebenso auf diesem Weg herzlich danke. Mit Bhaskar Chatterjee entwickelte sich im Anschluss eine mehrjährige Zusammenarbeit. Seine Untersuchungen bildeten eine essentielle Plattform. Ich wünsche Bhaskar Chatterjee das aller Beste für seinen weiteren Werdegang und meinen erkenntlichsten Dank. Ebenso gilt Philipp Kotter meiner vollster Dank für die gemeinsamen Untersuchungen im Bereich der Strukturdynamik. Die Unterhaltenungen mit ihm waren stets produktiv und furchtbar für die Durchführung dieser Arbeit. Da während dieser Arbeit mehrwöchige Messungen durchgeführt wurden, möchte ich ebenso dem Prüfstandsteam für ihre Unterstützung und Geduld danken. Hervorzuheben sind hierbei Florian Dräger und Rogelio Villanueva Porras, welche mich in diesen Tätigkeiten jeder Zeit unterstützen. Auf diesem Weg möchte ich ebenso meinen System- und Softwarekollegen danken, welche mich auch in herausfordernden Zeiten nicht aufkündigten.

Zur Durchführung solch einer Arbeit gehört einerseits das berufliche Umfeld. Andererseits trägt das private Umfeld mindestens das selbe Gewicht. Zunächst möchte ich meinen Eltern Hubert und Ursula Bischof für ihre geistige und finanzielle Unterstützung im Studium und darüber hinaus danken. Ebenso danke ich meinen Geschwistern Christian und Angelika Bischof sowie deren Partnern und deren Familien für die stetige Ermunterung und Aufheiterung auch in herausforderden Zeiten hindurch. Hinzu gilt mein herzlichster Dank meiner Freundin Sophie Eberhardt, welche mich in der Umsetzung dieser Arbeit jeder Zeit unterstütze, mir genügend Muse verlieh und mir in dieser Zeit ein absehbares Ziel vermittelte. Abschließend danke ich meinen Freunden für den Rückhalt und die stetige Ermunterung.

Ludwigsburg, den 07.11.2018

Deutsche Kurzfassung

In dieser Arbeit werden Modellierungsmethoden für das akustische Verhalten von Asynchronmaschinen, sowie Optimierungsansätze zur Regelung des akustischen Verhaltens vorgestellt. Zunächst werden in dieser Arbeit Ansätze zur effizienten Modellierung von Oberwellenerscheinungen der Elektromagnetik von Asynchronmaschinen aufgezeigt und analysiert. Hierbei fokussiert sich diese Arbeit zunächst auf die Entwicklung effizienter geometrischer Modelle zur Berechnung der Elektromagnetik von Asynchronmaschinen, welche größtenteils auf analytischen Gleichungen basieren und über numerische Methoden in einem erweiterten Umfeld eingesetzt werden. Im Anschluss daran wird eine Methode zur akustischen Berechnung, der aus der elektromagnetischen Kraftanregung entstehenden Vibrationen auf der Oberfläche des Gehäuses der elektrischen Maschine, vorgestellt. Der Ansatz zur Berechnung der Oberflächenvibrationen basiert auf einem Systemsimulations- und einem Schwingungssyntheseansatz. Das vorgestellte Verfahren bietet die Möglichkeit umfangreiche und realitätsnahe Modelle darzustellen und kann somit in starken Maßen zur optimierten Auslegung von Asynchron-, sowie vieler weiterer Arten von elektrischen Antriebsystemen beitragen. Das Verfahren besteht aus einem zweistufigen Prozess, wobei zwischen einer Offline-Berechnung, zu verstehen wie eine Vorausberechnung zur Modellparametrierung und zur Berechnung des elektromagnetischen Verhaltens, sowie einer Online-Berechnung, welche die eigentliche Akustikberechnung beinhaltet, unterschieden wird. In der Online-Berechnung werden die an die Systemsimulation gekoppelte elektromagnetische Kraftanregung auf das vorparametrierte Stator-, sowie das Gehäusemodell aufgebracht und anschließend in der Schwingungssynthese in Vibrationen umgerechnet. Durch die Kopplung der Systemsimulation an die elektromagnetische Kraftanregung können ebenso Regelungs- und Sensoreinflüsse, sowie Raumharmonische und Umrichter Schaltfrequenz Ordnungen analysiert und auf Grund der effizienten Berechnungsweise optimiert werden. Zusätzlich zur effizienten Berechnung der elektromagnetischen Oberwellen- und Oberschwingungserscheinungen kommt die effiziente Synthese über den Superpositionsansatz hinzu, welche die Vibrationsberechnung mit Hilfe von Kraft- und Strukturformen umsetzt. Der Schwingungssyntheseprozess ist für alle Arten von Antriebstopologien und Bauformen geeignet. Die Erkenntnisse dieser Berechnung werden anschließend zur Optimierung der Regelung des elektrischen Antriebs genutzt. Hierbei werden zwei Verfahren vorgestellt, welche sich ausschließlich auf die Änderungen der Software und deren Parametrierung des elektrischen Antriebstrangs beziehen. Zum einen wird in dieser Arbeit ein Verfahren vorgestellt, welches mit Hilfe der Änderung des Betriebspunktes bei gleichbleibendem Drehmoment und gleichbleibender Drehzahl das akustische Verhalten verbessert. In diesem Verfahren werden somit die Stromamplituden, Phasenlage und Stator Grundschwingungsfrequenz der Asynchronmaschine zur Verbesserung der Akustik des Antriebstrangs optimiert. Ein weiteres Verfahren beschreibt die Injektion von Spannungsimpulsen zur Erzeugung von harmonischen Stromschwingungen, welche der Anregung von störenden Vibrationen entgegenwirken und so zur Reduktion und Optimierung des akustischen Verhaltens der Asynchronmaschine und somit des elektrischen Antriebstrangs beitragen.

Contents

1 Introduction   **1**

2 Fundamentals   **7**
2.1 Noise Emissions and Noise Sources in Electrical Drives 7
2.2 Terms and Definitions for Machine Acoustics 9
    2.2.1 Definition of Waves and Vibrations 9
    2.2.2 Signal Processing and Acoustic Analysis 10
2.3 Inverter and Machine Control 17
    2.3.1 Basics in Machine Control 17
    2.3.2 Considered Inverter Modulation Schemes 18

3 Electromagnetic Modeling of Induction Machines   **23**
3.1 Sequential Electromagnetic Simulation 25
    3.1.1 Current Calculation 25
    3.1.2 Force Model 31
3.2 Integrated Electromagnetic Simulation 52
    3.2.1 Modular Air-gap Flux Approach 54
    3.2.2 Identification of Occurring Orders 61
3.3 Validation and Comparison 64
    3.3.1 Validation with Simulations 64
    3.3.2 Validation with Measurements 86
3.4 Conclusion 88

4 Structural Dynamic Modeling   **91**
4.1 Radial Air-gap Force Decomposition 92
4.2 Stator Tooth Force 94
4.3 Structural Response Analysis 96
4.4 Surface Displacement Synthesis 97
4.5 Validation and Comparison 98
    4.5.1 Comparison of Tooth Force Simulations 99
    4.5.2 Comparison of Structural Transfer Functions (TFs) 100
    4.5.3 Resultant Structure and Force Shapes 101
    4.5.4 Validation with Surface Velocity Measurements 104
4.6 Conclusion 106

Contents

5 Operating Point Adaption   **107**
5.1 Concept for Operating Point Adaption 108
5.2 Acoustic Measurements 112
5.3 Calculation of the Operating Points 116
    5.3.1 Operating Points for Best Efficiency 116
    5.3.2 Operating Points for Best NVH-behavior 116
5.4 Validation of the Optimization 119
5.5 Side Effects of Operating Point Adaption 119
    5.5.1 Stator Current 119
    5.5.2 Rotor Current 120
    5.5.3 Electromagnetic Torque 120
5.6 Conclusion 122

6 Elimination of Harmonics   **123**
6.1 Open Loop System: Control of Eccentric Rotor Position 125
    6.1.1 Definition of Eccentric Rotor Positions 125
    6.1.2 Principle of Rotor Eccentricity Elimination 125
6.2 Closed Loop System: Control on Saturation and Slotting Harmonics 126
    6.2.1 Repetitive Controller in d-q-axis Coordinate System 126
    6.2.2 Optimization Method with Minimization Algorithm 127
    6.2.3 Proportional-integral (PI) - based Harmonic Controller 128
    6.2.4 Ability of Harmonic Controllers 133
    6.2.5 Principle of Saturation and Slotting Harmonics Elimination 134
6.3 Validation and Comparison 136
    6.3.1 Control on Eccentricity Harmonics 136
    6.3.2 Control on Saturation and Slotting Harmonics 137
6.4 Conclusion 146

7 Compendium and Outlook   **149**

Appendix   **153**
A Induction Machine Drive Data 153
B Signal Flow Chart 154
C Comparison of the Electromagnetic Simulation Methods 155
D Park Transformation 156
E Clarke Transformation 156
F Derivation of Occurring Spatial Orders 156
G Comparison of the Electromagnetic Simulation Results 158
H Constant Tooth Force Approach (tooth) 158
I Modal Analysis 159
J Harmonic Analysis 159
K Modal Superposition 160

Contents

L   Force Response Calculation 162
M   Extension for Sound Power Calculation 163

List of Figures   **167**

List of Tables   **173**

Nomenclature   **175**

Acronyms   **183**

Bibliography   **187**

1 Introduction

With the beginning of the age of electromobility, the automotive manufacturers and their suppliers face new challenges. Customers think differently and demand new mobility solutions. This changes previous requirements for the construction of automobiles. The components are different, as well as the way of vehicle manufacturing and the vehicle system itself. The electrical drive solutions influence the perception of the Electrical Vehicle (EV), based on the drive performance, fuel feed, as well as the sound intensity of the drive. These facts lead to new requirements for electrical vehicles, which entail special challenges regarding acoustic requirements and new analysis opportunities. Before starting the electrical drive system and analysis, it is necessary to understand the construction of an EV.

The EV example in Fig. 1.1, shows that the combustion engine in the front area of a vehicle is changed to an electrical machine (3). High voltage wires (orange) connect the electrical machine (3) with the inverter (2). The inverter (2) is connected to the battery (1), which replaces the gas tank. For the inverter control system, the 12V battery is connected to the inverter with the low power supply cable (black). The EV needs to have an electrical brake booster (4), which is connected to the mechanical brakes via hydraulic tubes (turquoise). Finally, the electrical control unit (5) controls and communicates with the electrical drive and all of its system components (magenta).

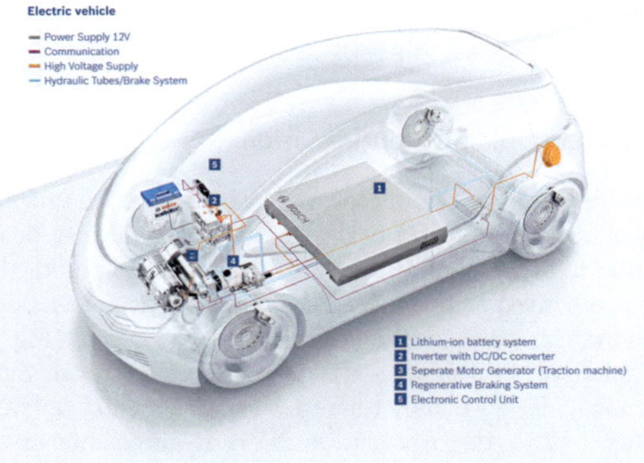

Figure 1.1: Construction of an Electrical Vehicle (EV)

Market potential and importance of the EVs

There are advantages as well as disadvantages in electromobility solutions. From the economic point of view, the forecasts for global sales and worldwide inventory development of EVs, diagrammed in Fig. 1.2, shows a high market potential. In these forecasts, the market for electromobility will increase drastically and fast. Till the year of 2030, there shall be mainly EVs or plug in EVs moving in our streets. The rapid expansion of EVs also entails risks. The countries' electrical energy system and infrastructure need to be adapted to a higher demand in electrical power. However in the long run, electromobility involves lower operational emissions, lower noise emissions and lower local particulate matters [Ele12]. This in turn can also help countries achieving their climate target. In addition, the production of the EVs produce more emission in comparison to conventional vehicles, highlighting the battery production [Sch14].

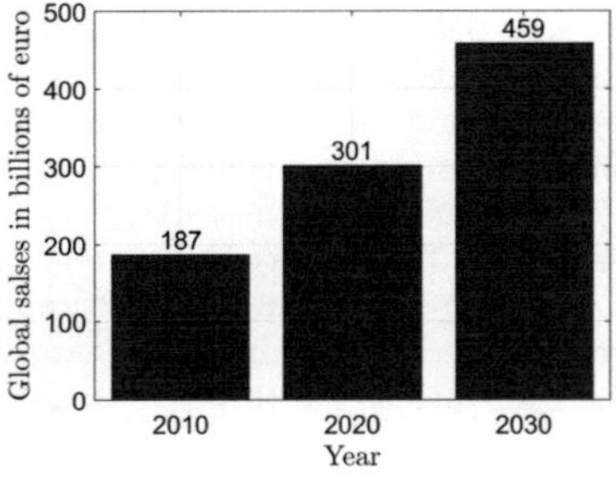

(a) Forecast of global sales of electric vehicles from 2010 to 2030 [McK11]

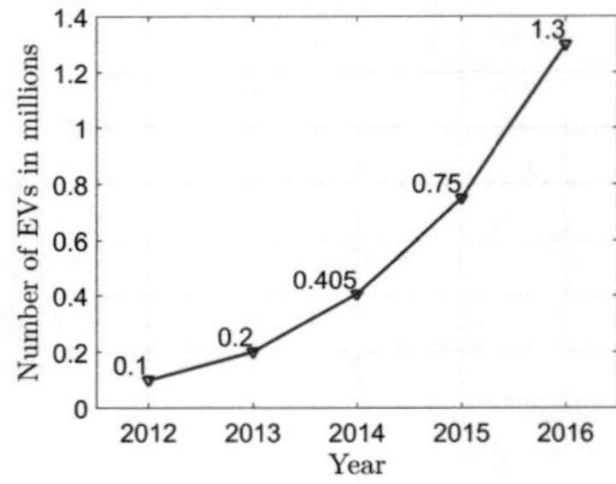

(b) Worldwide inventory development of electric cars in the years 2012 to 2016 [ZSW16]

Figure 1.2: Statistics of electrical vehicles

Electrical drives as multi-objective optimization

To fulfill the customer needs for acceleration, end speed, operating range and costs as well as comfort, the electrical drive and especially the electrical machine can be designed in a multi-objective optimization [NZE05; Ram+16]. With this design method, the priorities for the design characteristics are given to the optimizer in order to achieve a fast, cost-effective and highly requirements-matching electrical machine and drives design (Fig. 1.3). The technical design of the electrical drive often shows less focus on Noise, Vibration and Harshness (NVH) optimization in the early design phase, whereas the fundamental operational requirements are prioritized higher. With these priorities, NVH requirements are often not fulfilled. Thus, this work aims to simulate the harmonics in early design phases and shows how to change and optimize the control system of the inverter and electrical machine to achieve a cost-effective software solution fitting all customers needs.

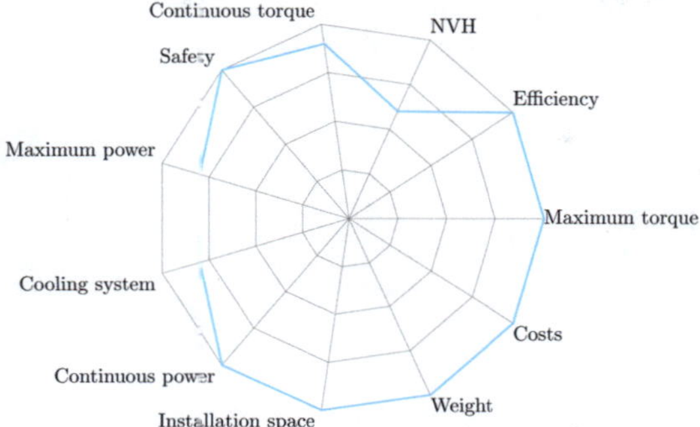

Figure 1.3: Example for priority of design characteristics of multi-objective optimization

Application of induction machine (asynchronous machine) drives

In former times, with a few minor exceptions, induction machines were mainly used in industrial applications, because of their reliability, low-maintenance and low costs [MP06; MVP08]. With the beginning of hybrid electromobility, permanently excited synchronous machines (PSMs) were applied. With the age of full electric vehicles a new discussion about the application of PSM and induction machine drives started. The difference of these electrical machines is mainly on the rotor side. Whereas PSM have permanent magnets in its rotor, the induction machine, has a squirrel cage, which in traction applications is built with copper bars to improve efficiency. These facts have a direct influence on efficiency and performance. The PSM shows a higher average efficiency, where in the partial load area, especially in efficiency cycles, the induction machine can partially achieve higher efficiencies. PSM drive systems show higher torque densities, but are more critical with the demagnetization of their magnets when heating up. For high performance applications with bigger machines, the induction machine can show a better average performance, because of the scaling effects of the power losses and its control strategy in the field weakening area. Induction machines also have advantages for the safety system, where the induced voltage and with it the induction machine drive can be turned off. The squirrel cage induction machine drives dispense with permanent magnets. Furthermore, it reduces costs and the consumption of toxic or rare materials. With all of this information, the induction machine is still an alternative to the PSM, which is shown in the latest electrical vehicle applications [Rip07; Lam18].

Causality chain from voltage excitation to acoustic noises

Getting more and more into the detailed analysis of electrical machines and drives, as well as the acoustics of electrical drives, the causality chain from the voltage excitation to the acoustic noises can be diagrammed in a rough scheme, shown in Fig. 1.4 [vdGie11; Boe13; Gar+97; VBG92; GWC06], and is inspired by the construction of this work.

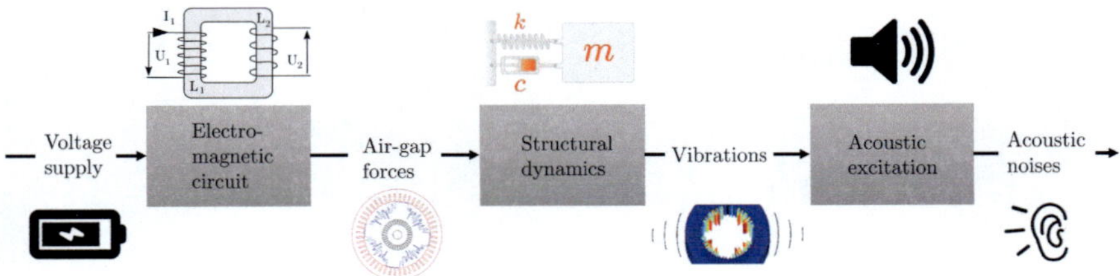

Figure 1.4: Causality chain from voltage excitation to acoustic noises

In chapter 2, the fundamentals for the acoustic analysis and signals processing as well as the inverter and machine control are explained. In this case, the inverter is the voltage supply, feeding the electromagnetic circuit, the induction machine, with voltage.

In chapter 3, the electromagnetic modeling of induction machines, is explained how the electromagnetic circuit creates air-gap forces. The electromagnetic circuit can be seen as a load drawing currents from the voltage source, creating a magnetic flux density. The magnetic flux density and subsequently the electromagnetic forces are creating tangential forces and radial forces directed to the stator and rotor, which produce electromagnetic torque and noise excitation.

These air-gap forces directed to the stator are described in chapter 4 on structural dynamic modeling. Projecting the forces on the stator and calculating the surface vibration response based on a structural dynamics model, the acoustic noises based on the surface vibration are computed.

After the calculation of the surface vibration, the electrical drive can be optimized for its NVH behavior with the findings from the previous chapters: The first approach is to change the electrical machine in its operating point, shown in chapter 5. This chapter shows the dependency of the machines operating strategy for best efficiency in conflict to best NVH behavior.
The second approach concerns to the elimination of harmonics in the currents and surface velocity, shown in chapter 6, injecting voltage harmonics in an open and closed looped system as well as showing the NVH- advantages and disadvantages.

Approach for efficient NVH simulations of induction machines

In order to optimize the analysis process for NVH simulations, the calculation path can be split up into two parts, shown in Fig. 1.5. The first part is the so-called 'offline' path, which can be seen as a pre-calculation, including the modal analysis of the mechanical structural dynamics and condensing it into a processable modal force matrix. After this pre-calculation, the 'online' part can be performed starting with the system simulation, calculating the inverter timings and currents for the electromagnetic model. This model includes the air-gap forces, which are used for estimating the air-gap forces and for the NVH-synthesis, where the structural dynamic are coupled to the electromagnetic [Kot+17b; Boe13].

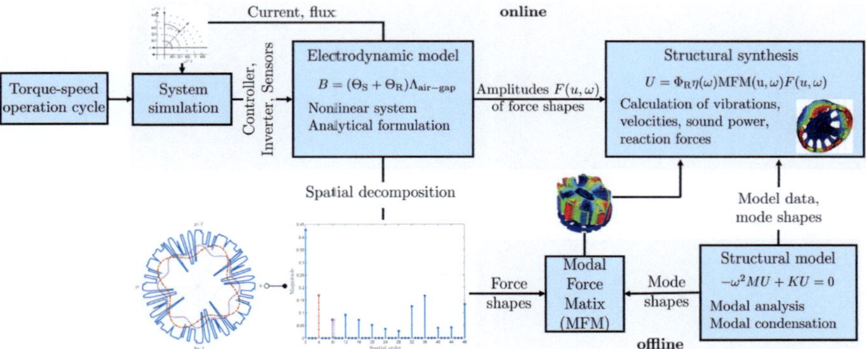

Figure 1.5: Simulation path for efficient NVH simulations (based on [Kot+17b; Boe13])

Introduction of the investigated electrical drive

The investigated electrical drive consists of an insulated-gate bipolar transistor (IGBT) inverter and a three-phase squirrel cage induction machine with aluminum, unskewed rotor bars and a distributed, zoned stator winding (App. A). The voltage level of the electrical drive is 180V to 250V with a maximum phase current of 500A. With the machines pole pair number of two, the electrical drive can provide a maximum torque of 133Nm and a maximum power of 70kW, where the power decreases to a continuous power of 35kW. The maximum speed of the machine is limited to 18000rpm with an axial length of 112mm and a stator outer diameter of 166mm. The induction machine has a mechanical one-sided air-gap of 0.6mm with 48 stator and 36 rotor slots. The inverter is mainly used with a switching frequency of 10kHz and with a space vector pulse width modulation scheme.

For the detailed analysis of the harmonics excitation, in chapter 3 and 4, especially the pole pair number, the number of stator and rotor slots as well as the switching frequency in relation to the rotor speed are the region of interest. With the number of pole pairs the order of the air-gap force excitation can be directly deviated. In case of the investigated electrical drive, the 0th, 4th and 8th force excitation are the dominant spatial orders. With the multiples of the stator and rotor slots, the temporal harmonics of the spatial forces can be roughly estimated, where the one-sided air-gap gives a hint on the electromagnetic damping of the system. Finally, the 10kHz of inverter switching frequency, the voltage modulation scheme and the stator frequency depending side bands of the interaction between the inverter and the electrical machine complete the picture.

2 Fundamentals

Before starting with modeling the electromagnetic and structural-dynamic behavior of electrical drives in chapter 3 and chapter 4, it is necessary to understand the general terms and definitions of electrical drives acoustics. In section 2.1 the noise emissions of drives are explained and the noise sources of electrical drives are presented. Section 2.2 shows the basics of waves and vibrations, as well as the calculation path from vibrations on the surface of the electrical drive to its sound power radiation. The acoustic analysis and its signal processing are shown in section 2.2.2 in order to show the different analysis methods, that are used in electrical drives acoustic analysis. The last section in this chapter treats the inverter control schemes and shows a rough implementation of the machine control system.

2.1 Noise Emissions and Noise Sources in Electrical Drives

Noise emissions have a huge influence and are surrounding us in our daily life [FSD12]. Especially noises in public have a huge influence on our sense of well-being and and cannot be directly influenced by oneself. One of the potential sources of these noises are vehicle. In particular, drives influence these noise emissions in public roads.

Too silent for pedestrians

Humans are familiar with the sounds of conventional vehicles through the daily use of cars. New generations of drives, especially electrical drives, change the picture of surrounding sounds. In fact, if pedestrians are distracted, electrical traction drives at low speeds can be too silent for their perception. Thus, pedestrians are terrified of EVs moving behind them or suddenly crossing their ways. To prevent these situations and to protect pedestrians, the European Union (EU) and the United States (US) view the possibility of new regulations for electromobility [Mas14; Com10]. One obvious solution is reviewed adding a sound generator to the electrical vehicle for the time of familiarization with the electrical vehicles and their electrical drives.

Not noisy but annoying

Comparing the noise emissions of electrical drives against conventional drives, the electrical drives have the potential of being less noisy, but also of being annoying in particular [BTM14]. The electrical machines are operating and inclining to tonal noises at higher frequencies. The conventional combustion engines have a more broadband noise and operating in the lower frequency range [Kön07]. Figure 2.1 shows the noise emissions of the vehicles components and their valence. It is shown, that the sound pressure of the vehicle should be as close as possible to the blue curve. The horizontal lines show the frequency components of the vehicle. In addition the electrical drive is diagrammed, in order to show the increased annoyance starting at around 1000Hz. The amplitude of the signals can also have a major influence, but in the test of [Kön07], noises don't have to be particularly loud to be annoying.

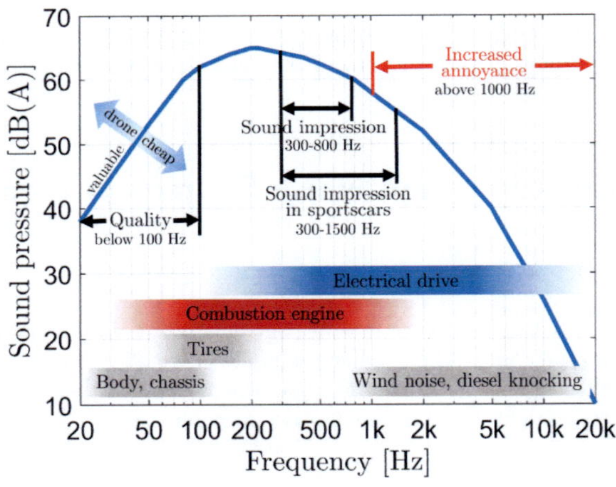

Figure 2.1: Noise emission of automobiles (based on [Kön07])

Noise sources in electrical drives

In order to understand the different noise sources of electrical drives, it is necessary to distinguish between noises of electromagnetic origin, mechanical origin, the interaction of mechanical and electrical origin, and the aerodynamic origin [Boe13; GWC06], shown in Fig. 2.2. Highlighting the mechanical and electromagnetic source, because these effects are only there in case of a non-ideal realistic electrical machine and have their origin normally in the manufacturing of the machine, affecting the electromagnetic and mechanical noises at the same time. Where the mechanical harmonics mainly include the bearing and friction

noises and the electromagnetic harmonics mainly include the radial forces, torque ripple and inverter switching. The transmission path of noises can be very different, where the main paths are the stator, housing and auxiliaries. From these transmission paths the structural borne noises are transmitted to accustic noises, also including the aerodynamic noises from cooling and slotting effects. Noises of aerodynamic harmonics and mechanical harmonics, not effecting the electromagnetic behavior, are mainly not in scope of this work.

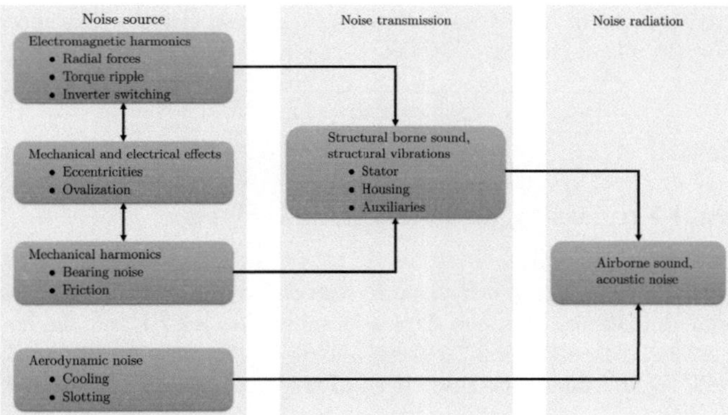

Figure 2.2: Noise sources in electrical drives

2.2 Terms and Definitions for Machine Acoustics

This section introduces the mathematical and physical expressions, which are needed for the general understandings of machine acoustics. Subsection 2.2.1 defines the general terms of waves and vibrations. Subsection 2.2.2 outlines the signal processing tools for acoustic analysis.

2.2.1 Definition of Waves and Vibrations

Vibrations are defined as oscillations on mechanical equilibrium points. It can be expressed as shown in (2.1), with its amplitude $\hat{y}$, time t, angular frequency ω and temporal harmonic order n [MM04]. The temporal harmonic order n is related to the angular frequency ω, thus $|n| = 1$ represents the fundamental vibration [MP06].

$$y_n(t) = \hat{y}\sin(n\omega t), \text{ with } n \in \mathbb{Z} \tag{2.1}$$

Waves are defined with a temporal part $n\omega t$ and its spatial part $\nu k\alpha$. The sign of the spatial part defines the direction the wave moves. Waves are hereby, not like vibrations, oscillations on mechanical equilibrium points, shown in equation (2.2), with the spatial velocity k, the spatial points α and the spatial harmonic order ν. The spatial harmonic order ν is related to the spatial velocity k, thus $|\nu| = 1$ defines the fundamental wave with its temporal distribution parts n. A spatial wave can herewith oscillate temporal with order n, but keeps its spatial distribution ν. A special form of the wave is the standing wave, which occurs due to the contrary movement of two waves with the same amplitudes. This special form will play an important role in chapter 4 [MP06].

$$\boxed{y_{\mathrm{n},\nu}(\alpha,t) = \hat{y}\sin(n\omega t \pm \nu k\alpha), \text{ with } n,\, k \in \mathbb{Z}} \tag{2.2}$$

2.2.2 Signal Processing and Acoustic Analysis

In order to analyze the signal behavior of electrical drives, it's necessary to define the constraint and characteristics of the signals in subsection 2.2.2.1 and the transformation methods in 2.2.2.2. Subsection 2.2.2.3 shows the display formats the of fast Fourier transformations (FFTs) and finally the discretized filters in subsection 2.2.2.4, that are used in this work.

2.2.2.1 Signal Constraint and Characteristics

This subsection defines the requirements on the signal in time and space in order to transform it into its frequency domain [Boe13].

Periodicity

Time signals must have at least one period of a constant amplitude in order to transform into the frequency domain. Speed run-ups are performed, so that this condition is fulfilled in comparison to the relevant frequency. With this, quantities in the air-gap of the rotating induction machine are always periodic and stationary in one period.

Discretization

Time signals have to be sampled in order to discretize the signal and to analyze their behavior. The sampling of this signal has to occur in equidistant steps. Beforehand the time signal has to be filtered with at least half of the highest relevant signal frequency content [Nyq28]. The filter methods and their discretizition are shown in subsection 2.2.2.4.

Fourier transform

There are two transformation domains, that are required for analyzing machine acoustics. The temporal frequency domain (TFD) signals in frequency f, or angular frequency $\omega = 2\pi f$ are defined for the temporal time domain (TTD) signals in time t as, shown in equation (2.3).

$$f(t) \quad \circ\!\!\!-\!\!\!\bullet \quad F(\omega) \tag{2.3}$$

The spatial frequency domain (SFD) signals in u are defined in spatial time domain (STD) in α, shown in equation (2.4) [Eff00].

$$f(\alpha) \quad \circ\!\!\!-\!\!\!\bullet \quad F(u) \tag{2.4}$$

Fourier series approximation and convention

Fourier transformations are individual linear frequency components and can therefore be linearly added in series, shown in subsection 2.2.2.2. If a function $f(t)$ or $f(\alpha)$ is real in time domain, then the function in frequency domain $F(\omega)$ or $F(\alpha)$ is symmetric to its origin. Fourier coefficients for the harmonic order ν can be written as two real values, shown in (2.5a) or combined in one complex number, shown in (2.5b). In this work, complex quantities are underlined and displayed in peak values.

$$a_{\nu,\mathrm{n}} = \Re\{\underline{c}_{\nu,\mathrm{n}}\} = \underline{a}_{\nu,\mathrm{n}}, \; b_{\nu,\mathrm{n}} = \Im\{\underline{b}_{\nu,\mathrm{n}}\} = \underline{b}_{\nu,\mathrm{n}} \tag{2.5a}$$

$$\underline{c}_{\nu,\mathrm{n}} = a_{\nu,\mathrm{n}} \pm jb_{\nu,\mathrm{n}} \tag{2.5b}$$

2.2.2.2 Fourier Series

There are different formulas of Fourier series, that can be taken for signal analysis. This subsection gives an overview of the Fourier series considered in this work.

Amplitude-phase Fourier series

A signal is periodic with period P and sampled in equidistant steps N. The spectral composition of the signal can be written as,

$$f(\alpha) = \frac{a_0}{2} + \sum_{\nu=1}^{\mathrm{N}} A_\nu \sin\left(\frac{2\pi\nu\alpha}{P} + \varphi_\nu\right) \tag{2.6}$$

where,

$$A_\nu = \sqrt{a_\nu^2 + b_\nu^2} \text{ and } \varphi_\nu = \arccos\frac{a_\nu}{A_\nu} = \arcsin\frac{b_\nu}{A_\nu} = \mathrm{atan2}(b_\nu, a_\nu) \tag{2.7}$$

and the Fourier coefficients a_ν and b_ν are defined as,

$$a_\nu = \frac{1}{P}\int_{-P}^{P} f(\alpha)\cos(\nu\alpha)\mathrm{d}\alpha \text{ for } \nu \geq 0 \tag{2.8a}$$

$$b_\nu = \frac{1}{P}\int_{-P}^{P} f(\alpha)\sin(\nu\alpha)\mathrm{d}\alpha \text{ for } \nu \geq 1. \tag{2.8b}$$

Sine-cosine sequence Fourier series

The amplitude-phase Fourier series in (2.6) can be rewritten with the sine addition theorem in its sine-cosine sequence Fourier series as follows,

$$f(\alpha) = \frac{a_0}{2} + \sum_{\nu=1}^{N}\left(a_\nu \cos(\frac{2\pi\nu\alpha}{P}) + b_\nu \sin(\frac{2\pi\nu\alpha}{P})\right) \tag{2.9}$$

where,

$$a_0 = A_0,\ a_\nu = A_\nu \sin(\varphi_\nu) \text{ and } b_\nu = A_\nu \cos(\varphi_\nu). \tag{2.10}$$

Positive-negative sequence Fourier series

Lastly the sine-cosine sequence can be presented in the complex function, shown in (2.11).

$$\underline{f}(\alpha) = \sum_{\nu=-N}^{N} \underline{c}_\nu e^{j\frac{2\pi\nu\alpha}{P}} \tag{2.11}$$

where,

$$\underline{c}_\nu = \begin{cases} \frac{1}{2}(a_\nu - jb_\nu) & \text{for } \nu > 0 \\ \frac{1}{2}a_0 & \text{for } \nu = 0 \\ \frac{1}{2}(a_\nu + jb_\nu) & \text{for } \nu < 0. \end{cases} \tag{2.12}$$

Equation (2.11) can be rewritten as shown in (2.13).

$$\underline{f}(\alpha) = \sum_{\nu=0}^{N} \underline{c}_\nu e^{j\frac{2\pi\nu\alpha}{P}} + \sum_{\nu=1}^{N} \underline{c}_{-\nu}^{*} e^{-j\frac{2\pi\nu\alpha}{P}} \tag{2.13}$$

2-D Fourier series

As shown in subsection 2.2.2.1, there are two transformations that are needed in order to analyze machine acoustics, due to the decomposition of waves in time and space. This transformation can be done in 1-dimensional (1-D) steps. First the temporal time domain (TTD) or the spatial time domain (STD) is transformed into its respective frequency domain,

(TFD) or (SFD) system. Next, the remaining time domain system is transformed into its respective frequency domain, shown in Fig. 2.3.

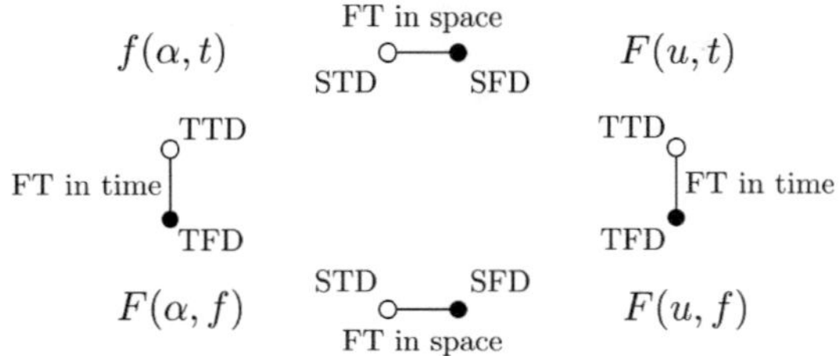

Figure 2.3: Transformation from TTD and STD to TFD and SFD

2.2.2.3 Display Formats of FFTs

This section introduces the display formats of fast Fourier transformations (FFTs). FFTs can be used to show the spectral components of a signal with at least one steady period. Depending on the transformation in spatial or temporal direction, either a FFT or a FFT based spectrum (Campbell diagram [Cam24]) should be displayed in order to proceed with the spectral analysis, that are mainly used in this work.

FFT: Magnitude – Order

The 2-dimensional (2-D) spectral analysis of a spatial or temporal signal is shown in the Fourier transform (FT) in Fig. 2.4. This transformation can be executed in magnitude over temporal or spatial order, that are belonging to the given frequency. In this work, the given frequency is always set to the fundamental electrical frequency of the signal.

FFT based Spectrum: Frequency – Magnitude – Time/Speed

The 3-dimensional (3-D) spectral analysis can be done with a FFT based spectrum [Cam24]. According to machine acoustics, a spatial signal is consistent in its spatial distribution and changing its temporal spectrum over time or rotor speed, this form of diagram is mainly used to visualize the temporal distribution of signal, shown in Fig. 2.4. The abscissa of this diagram shows the time or the rotor speed, which is referable to a given time step. The ordinate shows the frequency distribution and the colors show the magnitude of the signal component. The FFT based spectrum can be seen as a series of 2-D FFTs that fall into place for each time step or rotor speed, like a 2-D FFT over magnitude and frequency.

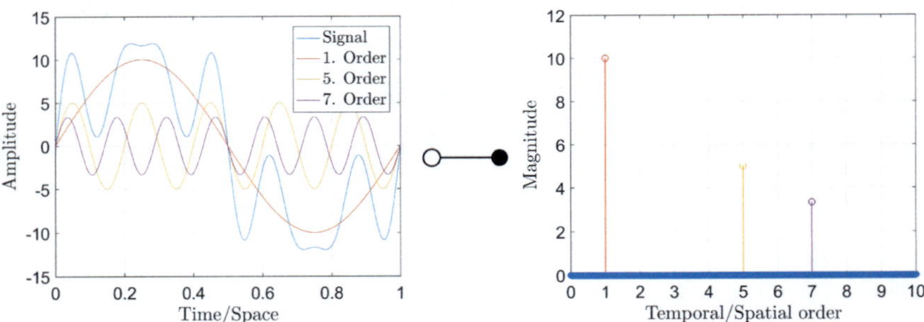

Figure 2.4: FT from TTD or STD to TFD or SFD

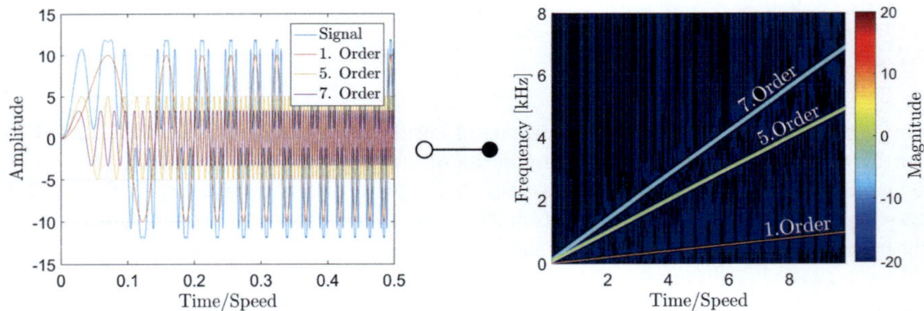

Figure 2.5: FFT based spectrum from TTD to TFD

The total harmonic distortion (THD)

Another method in order to characterize the harmonic content of a signal is to calculate the Total Harmonic Distortion (THD) [Cor15]. With this calculation, the amplitudes of specific orders can not be identified, but the harmonic distortion of the signals are easier comparable. Thus, the harmonic distortion is often used for measurements and signals relating to high harmonic content. Therefore the THD in % is defined with the fundamental amplitude A_1 and harmonic amplitudes A_n as,

$$THD = \sqrt{\frac{A_2^2 + \cdots + A_n^2}{A_1}} \tag{2.14}$$

2.2.2.4 Digital Filter Design

This subsection gives an overview of all kinds of filters, that are used in the next chapters. A general introduction into the topic of z-transformation, needed for the digital filter design is shown [Sop05; WT06]. There are two different specifications a filter should have [Ope06]:

Finite impulse response (FIR) filter

- Non recursive: Uses current and past input samples
- Has no analog equivalent
- More coefficients
  $\rightarrow$ more computation time
- Always stable

Infinite impulse response (IIR) filter

- Recursive: Uses current input, past input and output samples
- Has analog equivalent
- Less coefficients
  $\rightarrow$ less computation time
- Stability has to be proven

Because of the analog equivalent and less computation time, the infinite impulse response (IIR) filter design is chosen in this work [Sop05; WT06].

High-pass filter (HPF) and low-pass filter (LPF)

The first order analog high-pass filter (HPF) and LPF equations are shown in (2.15) and (2.16) with transfer parameter s and the angular cut off frequency $\omega_c = 2\pi f_c$, where the signal has a decrease of -3dB.

$$H_{\mathrm{HPF}}(s) = \frac{\frac{1}{\omega_c}s}{\frac{1}{\omega_c}s + 1} \qquad (2.15)$$

$$H_{\mathrm{LPF}}(s) = \frac{1}{\frac{1}{\omega_c}s + 1} \qquad (2.16)$$

The analog HPF and LPF equations are z-transformed into (2.17) and (2.18) with transfer parameter z and filter coefficient k_z [Hau08; Sta].

$$H_{\mathrm{HPF}}(z) = \frac{1 + k_{\mathrm{HPF}}}{2} \frac{z - 1}{z - k_{\mathrm{HPF}}} \qquad (2.17)$$

$$H_{\mathrm{LPF}}(z) = \frac{1 - k_{\mathrm{LPF}}}{2} \frac{z + 1}{z - k_{\mathrm{LPF}}} \qquad (2.18)$$

The coefficient k_z can be calculated for HPF and LPF with (2.19), with cut off frequency ω_c and the angular sampling frequency of $\omega_p = 2\pi f_p$. Both equations are stable for $|k_z| < 1$ and decrease at ω_c by -3dB.

$$k_{\mathrm{HPF}} = k_{\mathrm{LPF}} = \frac{1 - \sin\left(\frac{\omega_c}{\omega_p}\right)}{\cos\left(\frac{\omega_c}{\omega_p}\right)} \qquad (2.19)$$

The bode diagrams of the high-pass filter and low-pass filter are shown for a sampling frequency of $f_p = 5$ KHz and a cut off frequency of $f_c = 10$ Hz, in Fig. 2.6a and Fig. 2.6b. Due to the Nyquist-Shannon sampling theorem [Sha49], a perfect reconstruction is only

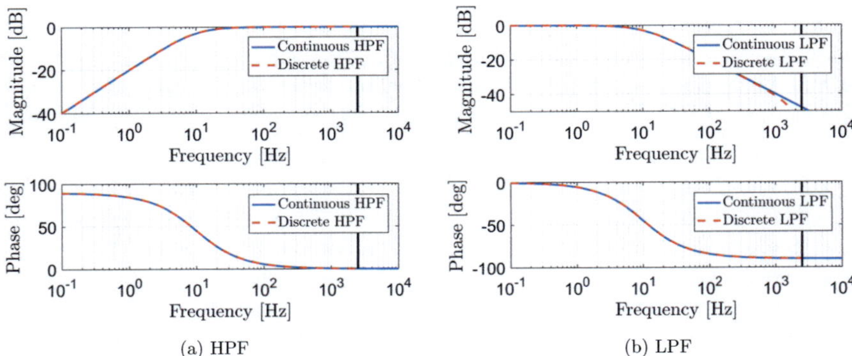

(a) HPF   (b) LPF

Figure 2.6: Bode diagram of (a) HPF and (b) LPF

guaranteed possible for a sample rate of half of the sampling frequency. With this fact, the discrete filters can only be shown till $\frac{f_\mathrm{P}}{2} = 2.5$ kHz.

Notch filter (NF) and peak filter (PF)

Notch filters (NFs) and peak filters (PFs) are used to filter a signal at one specific frequency. On the one hand, an analog notch filter (NF) with its angular notch frequency $\omega_\mathrm{N} = 2\pi f_\mathrm{N}$ is used in order to filter out the signal component with the frequency f_N, shown in (2.20) [SBS01; RJK94; LRO94]. On the other hand, an analog PF with its angular peak frequency $\omega_\mathrm{Pk} = 2\pi f_\mathrm{Pk}$ is used to filter out all signal components that haven't got the frequency f_Pk, shown in (2.21) [Zöl97; ST77; SH11]. With the fact of non ideal filters, frequencies around the notch or peak frequency are affected as well. The frequency bandwidth f_bw, or angular bandwidth frequency $\omega_\mathrm{bw} = 2\pi f_\mathrm{bw}$ that changes the width of influencing other frequencies can be set, but changes the damping of the aimed frequency as well.

$$H_\mathrm{NF}(s) = \frac{s^2 + \omega_\mathrm{N}^2}{s^2 + 2\omega_\mathrm{bw} + \omega_\mathrm{N}^2} \qquad (2.20) \qquad\qquad H_\mathrm{PF}(s) = \frac{2\omega_\mathrm{bw}s}{s^2 + 2\omega_\mathrm{bw} + \omega_\mathrm{P}^2} \qquad (2.21)$$

The z-transform of the notch filter and peak filter are shown in (2.22) and respectively (2.23) including their filter coefficients $k_\mathrm{NF,1}$ and $k_\mathrm{NF,2}$ for the notch filter as well as $k_\mathrm{PF,1}$ and $k_\mathrm{PF,2}$ for the peak filter.

$$H_\mathrm{NF}(z) = \frac{z^2 - 2k_\mathrm{NF,1}z + 1}{z^2 - 2k_\mathrm{NF,1}k_\mathrm{NF,2} + k_\mathrm{NF,2}^2} \qquad (2.22)$$

$$H_\mathrm{PF}(z) = \frac{(1 - k_\mathrm{PF,2}^2)z^2 - (1 - k_\mathrm{PF,2}^2)}{2(z^2 - 2k_\mathrm{PF,1}k_\mathrm{PF,2} + k_\mathrm{PF,2}^2)} \qquad (2.23)$$

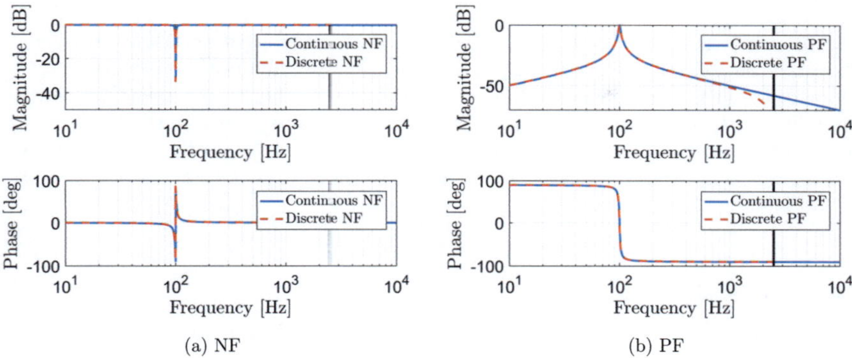

(a) NF   (b) PF

Figure 2.7: Bode diagram of a) NF and b) PF

The filter coefficients can be calculated with (2.24), (2.25) as well as (2.26), and are stable for $|k_{\mathrm{NF},1}| = |k_{\mathrm{NF},2}| < 1$ and $|k_{\mathrm{PF},1}| = |k_{\mathrm{PF},2}| < 1$.

$$k_{\mathrm{NF},1} = \cos(2\pi \frac{f_\mathrm{N}}{f_\mathrm{P}}) \qquad (2.24) \qquad\qquad k_{\mathrm{PF},1} = \cos(2\pi \frac{f_\mathrm{Pk}}{f_\mathrm{P}}) \qquad (2.25)$$

$$k_{\mathrm{NF},2} = k_{\mathrm{PF},2} = \frac{1 - \tan(2\pi \frac{f_\mathrm{bw}}{2 f_\mathrm{P}})}{1 + \tan(2\pi \frac{f_\mathrm{bw}}{2 f_\mathrm{P}})} \qquad (2.26)$$

The bode diagrams of the notch filter and peak filter are shown for a sampling frequency of $f_\mathrm{p} = 5$ KHz and a notch or peak frequency of $f_\mathrm{N} = f_\mathrm{Pk} = 100$ Hz, with a band width frequency of $f_\mathrm{bw} = 20$ Hz, in Fig. 2.7a and Fig. 2.7b. Due to the same Nyquist-Shannon sampling theorem [Sha49], as shown for HPF and LPF, the discrete filters can only be shown until $\frac{f_\mathrm{P}}{2} = 2.5$ kHz.

2.3 Inverter and Machine Control

This section gives a short overview of the inverter and machine control schemes. The first subsection 2.3.1 shows the fundamental machine controller design and its implementation. The second subsection 2.3.2 shows the considered inverter control schemes included in this work.

2.3.1 Basics in Machine Control

Machine controlling is an extensive field of controlling systems. Especially linear controllers, that are used for linearized operating points, are often used and are state of the art [Sch15; DPV11].

Proportional-Integral-Derivative PID controller

The proportional-integral-derivative (PID) controller is the most common controller for linearized control systems. Depending on the controlled system it can be easily parametrized and partly used as a proportional (P) or a proportional-integral (PI) controller. For machine controlling and its controlled system, the PID controller is used in order to eliminate an input error $e(t)$, shown in (2.27), with its control parameters K_{p}, as the proportional component, and K_{i}, as the integral component as well as K_{d}, as the derivative component.

$$y(t) = K_{\mathrm{p}}e(t) + K_{\mathrm{i}} \int_0^t e(\tau)\mathrm{d}\tau + K_{\mathrm{d}}\frac{\mathrm{d}e(t)}{\mathrm{d}t} \tag{2.27}$$

Cascade control

If there are different states in a control system and these control systems can be split up into smaller parts of the linear system, a cascade control can be used in order to control the complete system [Sch15; DPV11; SEM04]. In a cascade control, there is more than one controller cascading and interleaving for the subsystems and their transfer functions (TFs), shown in Fig. 2.8. In this example the inner loop is controlled with a PI controller, e.g. a current loop with TF_1 behavior, and the outer loop, e.g. a torque loop with TF_2 behavior, where the sampling of the inner loop has to be 10 times the outer loop. The cascade controlled signal flow chart of the controlled machine is shown in appendix B.

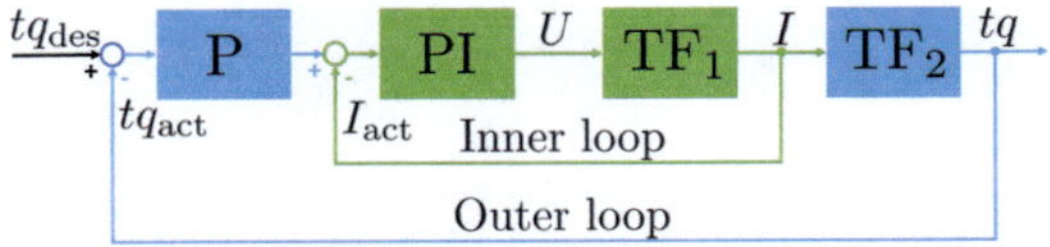

Figure 2.8: Schematic diagram of a cascade controlled system

2.3.2 Considered Inverter Modulation Schemes

In order to understand the inverter control schemes, it is necessary to go into detail with the fundamentals of the inverter control [Ken17; Sch12; Bus15]. Figure 2.9a shows the 3-phase inverter with the six IGBTs. These IGBTs can be simplified to switches, which can be turned on and off. Depending on the on timing of each transistor, specific voltages are created in order to reach the chosen operating point (OP) of the inverter and hence the electrical machine. The operating of this inverter topology is specified with a hexagon, that is divided up into six sectors and eight vectors, shown in Fig. 2.9b. Thus, an operating vector, and with it an operating voltage, can only be created with switching the vectors on and off. The number behind the edge of the hexagon defines the state of the IGBTs on the upper side. Zero stands for off, one stands for on. The lower side IGBTs are inverted, thus zero stands for on and an one stands for off. The timing of switching depends on the

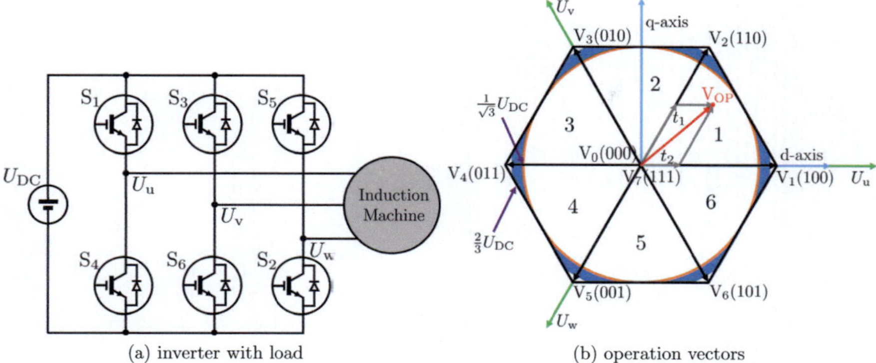

(a) inverter with load   (b) operation vectors

Figure 2.9: 3-phase (a) inverter with load and (b) operation vectors

inverter control scheme. Therefore different schemes lead to different voltages and harmonic excitation. This work considers the space vector pulse width (SVPWM), flat top pulse width (FTPWM) as well as over pulse width modulation (OVPWM) [Ken12; Ken13; Inf05; Sys03; ZW02; QD15].

Space vector pulse width modulation (SVPWM)

The space vector pulse width (SVPWM) is the most common used switching method of the inverter. It includes all eight vectors shown in Fig. 2.9b. A switching sequence example, including an example of the timings is shown in 2.10, with t_s as one PWM period. Figure 2.11a shows the switching of the IGBTs and the specific duty cycle for this inverter control scheme. The maximum output voltage of the fundamental of this procedure is limited by $\frac{1}{\sqrt{3}}U_{DC}$, diagrammed as an orange circle in Fig. 2.9b. The harmonic voltage components are shown in Fig. 2.11b, with a pulse width modulation (PWM) switching frequency f_{switch} of 10kHz. Orders that occur due to a linear

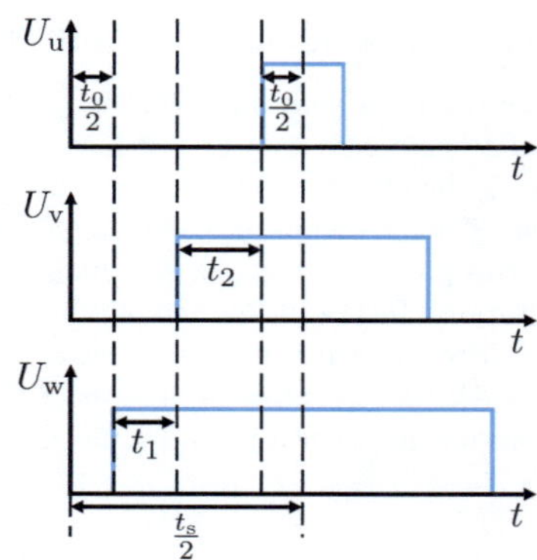

Figure 2.10: Switching sequence for SVPWM

load are defined in (2.28), with f_{elec} as the fundamental frequency of voltage signal and where f_{SVPWM} is never a multiple of 3 f_{elec} for star point connected machines [MUR06].

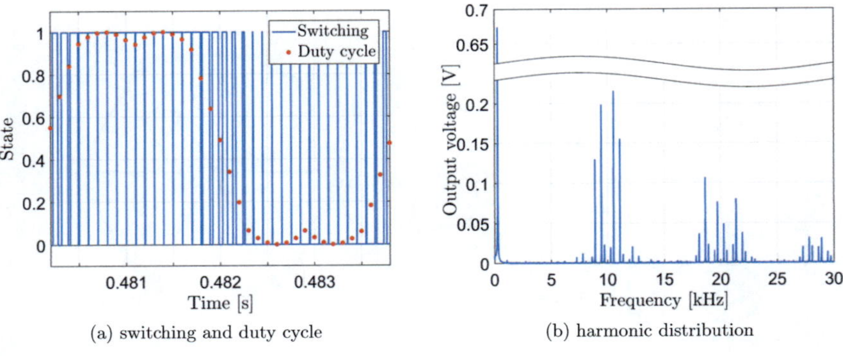

(a) switching and duty cycle

(b) harmonic distribution

Figure 2.11: SVPWM (a) switching and duty cycle and (b) harmonic distribution example

This inverter control scheme forms the basis for the flat top and over pulse width modulation procedures.

$$
f_{\mathrm{SVPWM}} = \begin{cases} f_{\mathrm{elec}} & \text{always} \\ k_{\mathrm{h}} f_{\mathrm{switch}} \pm k_{\mathrm{n}} f_{\mathrm{elec}} & \text{for } k_{\mathrm{h}} = 1,3,5,\dots \text{ and } k_{\mathrm{n}} = 2,4,6,\dots \\ k_{\mathrm{h}} f_{\mathrm{switch}} \pm k_{\mathrm{n}} f_{\mathrm{elec}} & \text{for } k_{\mathrm{h}} = 2,4,8,\dots \text{ and } k_{\mathrm{n}} = 1,5,7\dots \end{cases} \tag{2.28}
$$

Flat top pulse width modulation (FTPWM)

The flat top pulse width modulation (FTPWM) is a modified version of the SVPWM. This control scheme uses only one of zero vectors V_0 or V_7, that reduces switching losses and increases the efficiency of the inverter. Other vectors are timed differently, but are used as well. The maximum output voltage of this procedure is the same as SVPWM ($\frac{1}{\sqrt{3}}U_{\mathrm{DC}}$). Thus, this procedure reduces switching losses but increases the occurring voltage harmonics distribution. In Fig. 2.12a and 2.12b, the switching sequence is shown for using one zero vector. The characteristic duty cycle and the harmonics distribution is diagrammed in Fig. 2.13a and 2.13b. In comparison to the SVPWM, the duty cycle is getting longer. This causes more harmonics, shown in Fig. 2.12b. The occurring harmonics can also be calculated with (2.28).

Over pulse width modulation (OVPWM)

The over modulation pulse width modulation (OVPWM) is a method to increase the output voltage of the system over the $\frac{1}{\sqrt{3}}U_{\mathrm{DC}}$ limitation of SVPWM and FTPWM. It can be seen as a procedure like the FTPWM, but extending the vectors length, shown as a dark blue

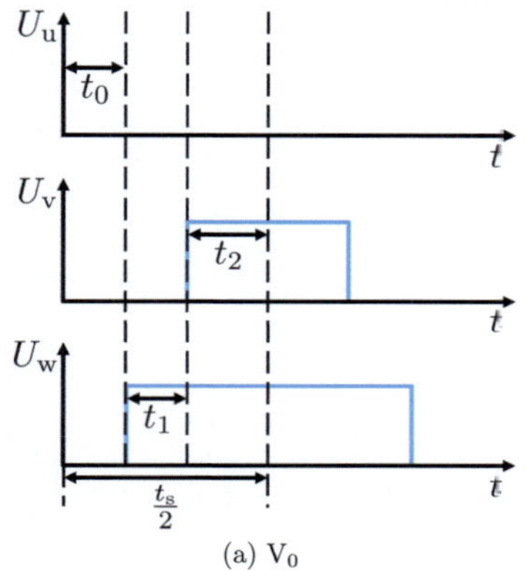

(a) V_0

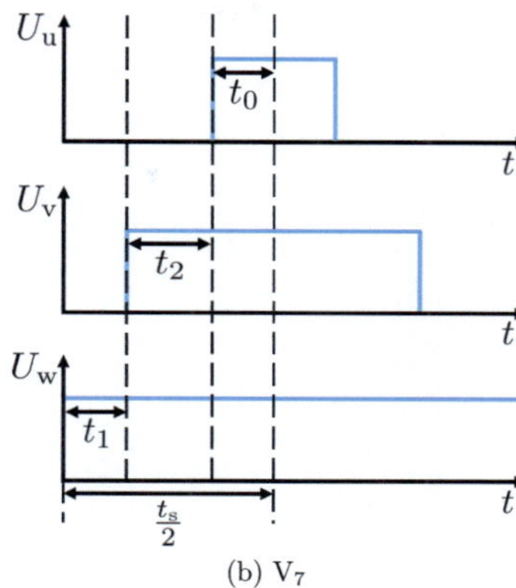

(b) V_7

Figure 2.12: Switching sequence for FTPWM using (a) V_0 and (b) V_7 as zero vector

surface in Fig. 2.9b. The maximum output voltage is limited to $\frac{2}{3}U_{\mathrm{DC}}$, but increasing the vectors till this voltage is equal to square wave mode. Thus, this inverter control scheme enables higher voltages and with it higher power, but increases the harmonics content of the voltage signal, due to lower sinusoidal voltage. The switching scheme and the characteristic duty cycle is shown for this OVPWM in Fig. 2.14a. The duty cycle is getting longer and longer and with it, the voltage gets more square components. This effect can also be seen in the harmonics distribution in Fig. 2.14b, where additional 5th, 7th,... harmonics are shown and caused by the switching scheme from Fig. 2.14a. The harmonics, that occur due to this inverter control scheme are defined in (2.29) for a linear load. Multiple of 3 f_{elec} also don't occur here, due to the star point connected machine.

$$f_{\mathrm{OVPWM}} = \begin{cases} f_{\mathrm{elec}}k_{\mathrm{n}} & \text{for } k_{\mathrm{n}} = 1,5,7 \\ k_{\mathrm{h}}f_{\mathrm{switch}} \pm k_{\mathrm{n}}f_{\mathrm{elec}} & \text{for } k_{\mathrm{h}} = 1,3,5,\ldots \text{ and } k_{\mathrm{n}} = 2,4,6,\ldots \\ k_{\mathrm{h}}f_{\mathrm{switch}} \pm k_{\mathrm{n}}f_{\mathrm{elec}} & \text{for } k_{\mathrm{h}} = 2,4,8,\ldots \text{ and } k_{\mathrm{n}} = 1,5,7\ldots \end{cases} \tag{2.29}$$

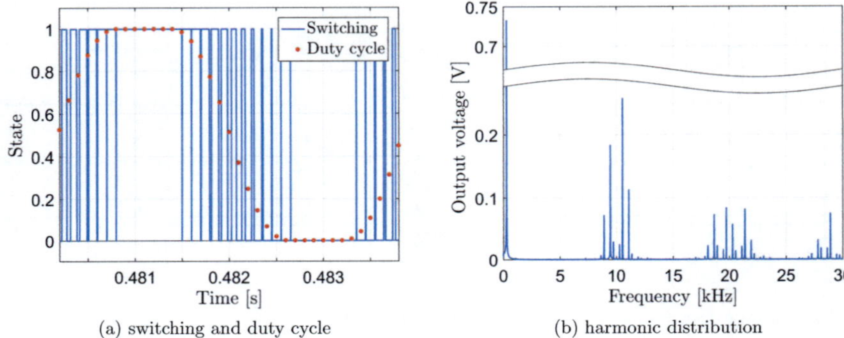

(a) switching and duty cycle

(b) harmonic distribution

Figure 2.13: FTPWM (a) switching and duty cycle and (b) harmonic distribution example

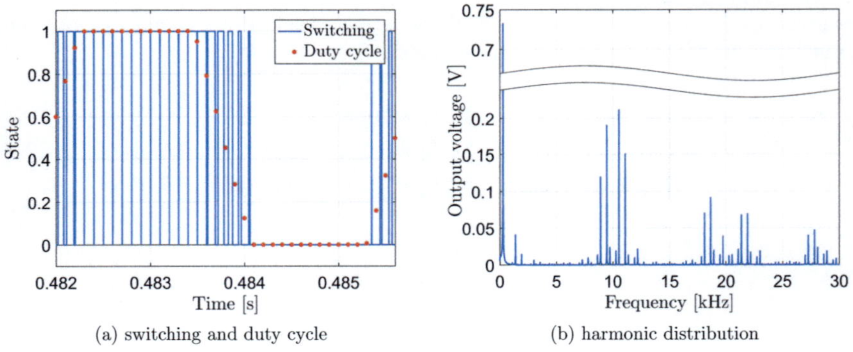

(a) switching and duty cycle

(b) harmonic distribution

Figure 2.14: OVPWM (a) switching and duty cycle and (b) harmonic distribution example

3 Electromagnetic Modeling of Induction Machines

This chapter shows the different implementation methods in order to calculate the electromagnetic behavior and the force excitation of the induction machine (App. A). With the introduction of the fundamental flux model in section 3.1.1, spatial harmonic components can be added sequentially, shown in section 3.1.2 to create a model including harmonics. In addition, based on section 3.1.2 a model is proposed including all spatial and temporal electromagnetic effects in one integrated model, shown in section 3.2. The simulation model is implemented in Mathworks Inc. MATrix LABobratory (MATLAB) & Simulink [Inc17b; Ang+11] and validated with the Finite Element method (FEM) software JSOL JMAG [JSO17].

To compare the simulation methods for electromagnetic induction machine calculations, in appendix C the geometrical permeance model structure is chosen, which is similar to a graphical reluctance model. The advantages of the permeance model are a shorter simulation time, causal relations as well as the coupling with inverter, controller and sensors. The permeance model can be improved regarding its accuracy, current and saturation calculation with applying the integrated electromagnetic simulation. Because of focusing on the simulation time and feeding the machine with a PWM-Inverter, it is prudent to use a geometrical model. In order to use the geometrical model, as shown in table C.1, assumptions have to be made as follows:

1. The permeability of vacuum μ_0 is much smaller than the permeability of iron μ_i ($\mu_0 << \mu_i$) [Bes08], and the magnetic flux lines enter almost perpendicularly into the iron, meaning the radial magnetic flux density B_r overshadows the tangential magnetic flux density B_t ($B_t << B_r$). The radial part of the Maxwell stress tensor σ_r can be approximated to [MP09; PLF12a; Klo98]:

$$\sigma_r = \frac{\mu_i - 1}{2\mu_i\mu_0}(B_r^2 + \mu_i B_t^2) \approx \frac{B_r^2}{2\mu_0}. \tag{3.1}$$

2. Current displacement, eddy currents and deep rotor bar effects are included in the fundamental induction machine parameters [MVP08; SA12; JST16; De 92]. Damping of higher harmonics is only included indirectly .

The fundamental equation that has to be solved in order to calculate the noise excitation of the electrical machine is the radial force equation, shown in (3.2), with $B_r(\alpha,t)$ as the radial magnetic flux density, μ_0 as the vacuum permeability and surface area A_{surf} [MVP08; Bes08; Bin12; BN09]. As the surface area A_{surf} and the vacuum permeability μ_0 are constant quantities, the radial magnetic flux density $B_{\text{air-gap}}(\alpha,t)$ has to be calculated. The equation describes the radial electromagnetic forces $F_r(\alpha,t)$ in the air-gap of the machine and affecting the stator, its structural dynamics system and with it the noise emissions.

$$F_r(\alpha,t) \approx \oiint \sigma_r \mathrm{d}A = \frac{B_{\text{air-gap}}^2(\alpha,t)}{2\mu_0} A_{\text{surf}} \tag{3.2}$$

With the knowledge of the mechanical surface area A_{surf}, the calculation of the radial force is mainly focused on the calculation of the radial magnetic flux density $B_{\text{air-gap}}(\alpha,t)$ in the air-gap of the induction machine. The radial magnetic flux density is defined as,

$$B_{\text{air-gap}}(\alpha,t) = \underbrace{(\Theta_s(\alpha,t) + \Theta_r(\alpha,t))}_{\Theta(\alpha,t)}\Lambda_{\text{air-gap}}(\alpha,t), \; B_{\text{air-gap}}(\alpha,t) \in \mathbb{R}. \tag{3.3}$$

with the stator and rotor magneto motive forces (mmfs), $\Theta_s(\alpha,t)$ and $\Theta_r(\alpha,t)$, as well as the air-gap permeance $\Lambda_{\text{air-gap}}(\alpha,t)$.

The procedures and equations in the following sections are split up into the spatial time domain (STD) and spatial frequency domain (SFD) in order to get a deeper understanding of the harmonics content of each component of the magnetic flux density. The calculation in the SFD allows a direct analysis of orders, which would only be possible with a post processing FFT and allows to work with steady functions. The spatial frequency domain SFD will also be important for the structural dynamic model and synthesis, shown in chapter 4.

The following chapters are divided into two different approaches, the sequential electromagnetic simulation, in section 3.1, and the integrated electromagnetic simulation, in section 3.2. For the sequential approach, the force model is calculated sequentially to the currents. The information from the calculated spatial magnetic flux density in the force model is not fed back into the current calculation. The current calculation in the sequential model is a derived model from the fundamental and harmonics behavior of the induction machine. This is the most common described procedure in literature. Different to this approach is the integrated electromagnetic simulation, which uses the information of the spatial magnetic flux density in order to calculate the stator and rotor currents with all of the included harmonic effects.

It is worth mentioning, that the NVH analysis in general includes more parameters, which can be important, like e.g. stator and rotor current harmonics and torque ripple. The machine itself can have a specific noise excitation via the radial path, but with installing the electrical drive in the car and coupling it on a transmission, harmonics that can be excited from the inverter itself from current harmonics or radial forces due to the torque ripple can also cause noise emissions. Because of these facts, this work also concerns electrical parameters and tries to show these effects as well.

3.1 Sequential Electromagnetic Simulation

The sequential electromagnetic simulation includes the temporal electromagnetic behavior, shown in Fig. 3.1, like the current calculation for the force model. The model structure is derived from the spatial electromagnetic behavior neglecting the temporal harmonic machine behavior. This temporal electromagnetic behavior is mainly described as the fundamental electromagnetic behavior of the induction machine. The force model calculates the spatial electromagnetic flux density in the air-gap of the induction machine and with it subsequently the radial machine forces.

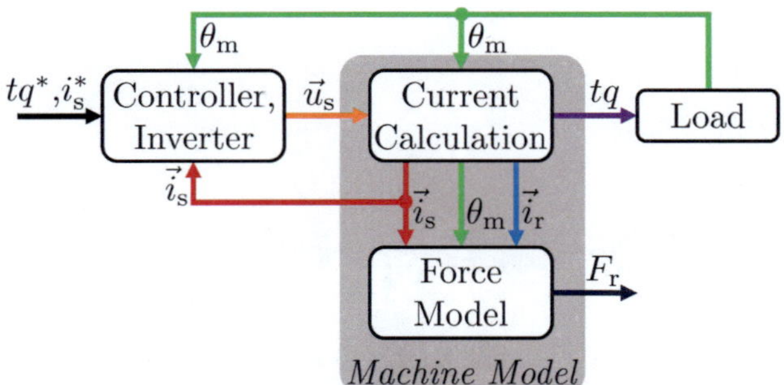

Figure 3.1: System integration of the sequential electromagnetic model

3.1.1 Current Calculation

The current calculation relates only to the temporal electromagnetic behavior, that is often described as the fundamental electromagnetic behavior of the induction machine [Bin12; MVP08]. This model structure is already derived from the spatial electromagnetic model and reduced by neglecting the spatial harmonic behavior. Harmonic effects can be partially added with extending the fundamental flux model, shown in subsection 3.1.1.1, and adding temporal harmonics of the inverter.

3.1.1.1 Fundamental Flux Model

The fundamental flux model is described in (3.4a) - (3.4e), with its resistance R, inductance L, flux linkage Ψ, voltage u, current i and the angular mechanical rotor frequency ω_m. The indices R, S, M stand for the rotor, stator and main, as well as σ for leakage. The equivalent circuit is shown in the Fig. 3.2, where capital letters describe the variable in the transformed system, lower case letters describe the variable in its own system.

As basis for the fundamental flux model, three different flux linkage orientations are taken into account. Based on the universal field-orientation approach, where $a = 1$ for main flux linkage orientation, $a = {}^{L_\mathrm{m}}/_{L_{\sigma,\mathrm{r}}}$ for rotor flux linkage orientation, as well as $a = {}^{L_{\sigma,\mathrm{s}}}/_{L_\mathrm{m}}$ for stator flux linkage orientation. Changing the field-orientation helps to simulate and control the system in an easier way, where usually the rotor flux linkage orientation is used. The parameter change can be calculated with (3.5a) - (3.5e) [DPV11; MVP08; Bin12].

$$\vec{u}_\mathrm{s} = R_\mathrm{s}\vec{i}_\mathrm{s} + \frac{\mathrm{d}\vec{\Psi}_\mathrm{S}}{\mathrm{d}t} \tag{3.4a}$$

$$\vec{\Psi}_\mathrm{S} = \vec{\Psi}_\mathrm{M} + L_{\sigma,\mathrm{S}}\vec{i}_\mathrm{s} \tag{3.4b}$$

$$\vec{\Psi}_\mathrm{R} = \vec{\Psi}_\mathrm{M} - L_{\sigma,\mathrm{R}}\vec{i}_\mathrm{R} \tag{3.4c}$$

$$\vec{\Psi}_\mathrm{M} = L_\mathrm{M}\vec{i}_\mathrm{Mag} \tag{3.4d}$$

$$\frac{\mathrm{d}\vec{\Psi}_\mathrm{R}}{\mathrm{d}t} = R_\mathrm{R}\vec{i}_\mathrm{R} + j\omega_\mathrm{m}\vec{\Psi}_\mathrm{R} \tag{3.4e}$$

$$L_{\sigma,\mathrm{S}} = L_{\sigma,\mathrm{s}} - aL_\mathrm{m} \tag{3.5a}$$

$$L_{\sigma,\mathrm{R}} = a^2 L_{\sigma,\mathrm{r}} - aL_\mathrm{m} \tag{3.5b}$$

$$L_\mathrm{M} = aL_\mathrm{m} \tag{3.5c}$$

$$\vec{i}_\mathrm{R} = \frac{\vec{i}_\mathrm{r}}{a} \tag{3.5d}$$

$$R_\mathrm{R} = a^2 R_\mathrm{r} \tag{3.5e}$$

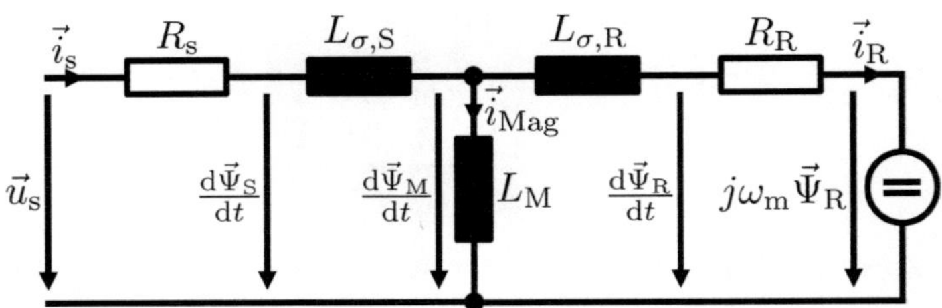

Figure 3.2: Equivalent circuit of the fundamental flux model

Fundamental flux model in d-q-coordinate system

The fundamental flux model can be described in the d-q-coordinate system, derived from the general induction machine equations (3.4a) - (3.4e). This derivation is operated by transforming the machine model into the rotor flux linkage oriented system [DPV11].

In order to use the steady d-q-coordinate system model, the three-phase U-V-W system has to be transformed with Park Transformation, shown in (D.1), including the rotor flux linkage angle θ_R.

Equations (3.6a) - (3.7c) define the d-q-axis model, shown in Fig. 3.3a and 3.3b, with d and q as indices depending on the axis, the angular stator frequency $\omega_s = 2\pi f_s$, slip frequency $\omega_{slip} = 2\pi f_{slip}$, machine pole pair number p and eletrical machine torque tq.

$$u_d = R_s i_d - \omega_s L_{\sigma,S} i_q + L_{\sigma,S}\frac{di_d}{dt} - \frac{d\Psi_M}{dt} \quad (3.6a)$$

$$u_q = R_s i_q + \omega_S L_{\sigma,S} i_d + L_{\sigma,S}\frac{di_q}{dt} + \omega_s \Psi_M \quad (3.6b)$$

$$\Psi_M = L_M(i_d - i_{d,R}) \quad (3.6c)$$

$$tq = 3p\Psi_M i_q \quad (3.7a)$$

$$\omega_{slip} = \frac{R_R i_q}{\Psi_M} \quad (3.7b)$$

$$\frac{d\Psi_M}{dt} = R_R i_{d,R} \quad (3.7c)$$

The equation (3.7c) shows a disadvantage of this representation. With reaching a steady operating point, the derivative of the rotor flux linkage and with it the rotor current $i_{d,R}$ becomes zero. With this fact, the rotor current can't be directly calculated with this system. But reaching steady operating points show a advantage for the controlling system especially for the flux observers.

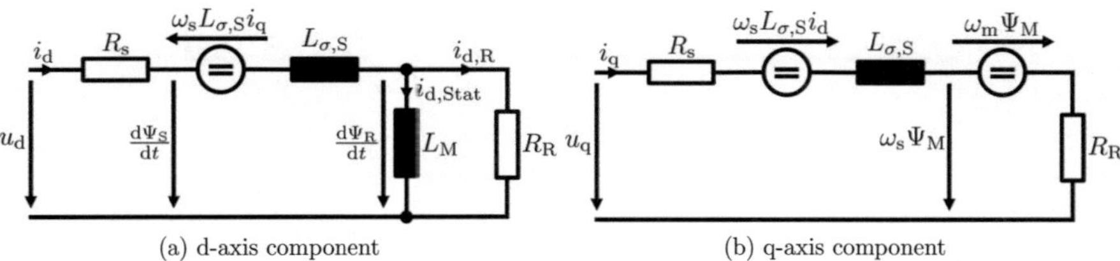

(a) d-axis component   (b) q-axis component

Figure 3.3: Fundamental flux model (a) d-axis component and (b) q-axis component

Finally, the d-q-axis currents are transformed back into the three-phase U-V-W system, using inverse Park Transformation, shown in (D.2) and depending on the rotor flux angle.

Fundamental flux model in α-β-coordinate system

Another procedure in order to simulate the transient electromagnetic effects, is to describe the fundamental equations of the induction machine with an ideal rotating transformer (IRTF) model [DPV11]. For this kind of model, the three-phase U-V-W-coordinate system has to be transformed into the fixed α-β-coordinate system, using Clark Transformation, as shown in (E.1).

The IRTF acts like an ideal air-gap and changes the rotating frequency of the stator and rotor side system depending on the mechanical angular frequency of the rotor ω_m, shown in Fig. 3.4. Quantities like the rotor current $\vec{i}_R$ or main flux linkage $\vec{\Psi}_m$ are transformed from the stator side to the rotor side or vice versa. The variables for the rotor side are displayed with indices xy. The fundamental induction machine equations are shown in (3.8a) - (3.8e).

$$\vec{u}_\mathrm{s} = R_\mathrm{s}\vec{i}_\mathrm{s} + \frac{\mathrm{d}\vec{\Psi}_\mathrm{S}}{\mathrm{d}t} \tag{3.8a}$$

$$\vec{\Psi}_\mathrm{S} = \vec{\Psi}_\mathrm{M} + L_{\sigma,\mathrm{S}}\vec{i}_\mathrm{s} \tag{3.8b}$$

$$\vec{\Psi}_\mathrm{R}^{\mathrm{xy}} = \vec{\Psi}_\mathrm{M}^{\mathrm{xy}} - L_{\sigma,\mathrm{R}}\vec{i}_\mathrm{R}^{\mathrm{xy}} \tag{3.8c}$$

$$\vec{\Psi}_\mathrm{M} = L_\mathrm{M}\vec{i}_\mathrm{Mag} \tag{3.8d}$$

$$\frac{\mathrm{d}\vec{\Psi}_\mathrm{R}^{\mathrm{xy}}}{\mathrm{d}t} = R_\mathrm{R}\vec{i}_\mathrm{R}^{\mathrm{xy}} \tag{3.8e}$$

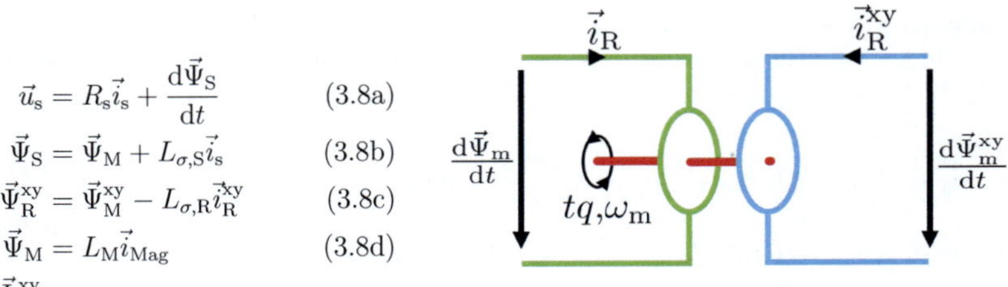

Figure 3.4: Schematic diagram of the ideal rotating transformer (IRTF)

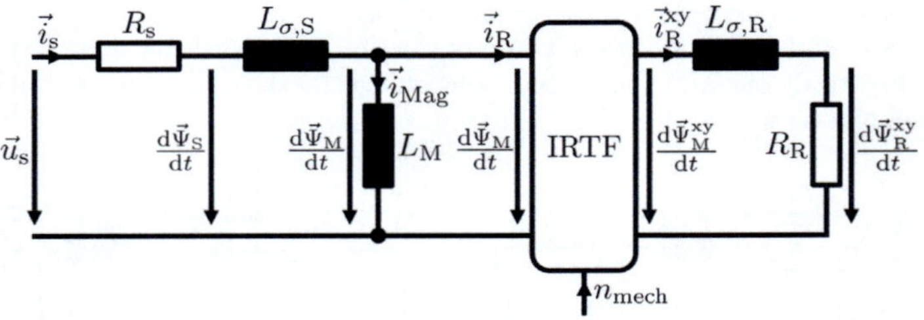

Figure 3.5: IRTF fundamental flux model (based on [DPV11])

After the calculation in α-β-coordinates, the currents have to be transformed back into the three-phase U-V-W-system with the inverse Clark Transformation, shown in (E.2).

3.1.1.2 Saturation of the Fundamental Flux Model

In order to ensure a valid operation in the entire operating range, the saturation behavior of the induction machine has to be considered. For this case, the parametrization of the models or the model structure has to be changed fto include the fundamental saturation behavior.

As the temporal electromagnetic models are parametrized from the steady state equations [DPV11; MVP08; Bin12], the model parameters can be changed for machine saturation. These quantities can also be saved as matrices in look-up-tables, depending on the operating point of the machine. This procedure is valid for steady operating points and gives a rough estimation in transients.

A more appropriate method is to compare the absolute values of the flux linkages with a B-H-curve of the similar machine material curve. This will limit the fluxes depending on the level of the flux linkages and is comparable to the fundamental flux saturation.

3.1.1.3 Extension of the Fundamental Model

In order to include harmonics that are induced from the spatial stator harmonics ν to the rotor side can be extended by setting additional rotor circuits in parallel to the fundamental rotor circuit (Fig. 3.6) [DPV11; MVP08; Bin12; Sei92; GWC06; Lac05]. This calculation includes additional information of the harmonic rotor leakage inductance $L_{\sigma,\mathrm{R}}^{\nu}$, main inductance L_{M}^{ν}, resistance R_{R}^{ν}, as well as the harmonic slip s_{ν} as,

$$s_{\nu} = 1 - \nu(1 - s). \tag{3.9}$$

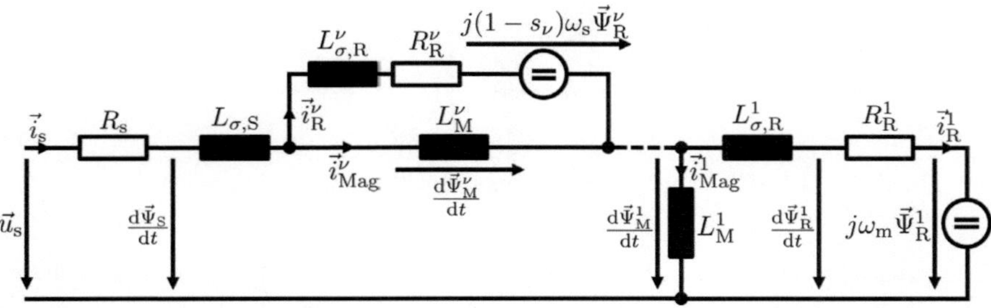

Figure 3.6: Fundamental flux model with rotor harmonics extension

The extended equations, shown in (3.10a) - (3.10c), can be used for steady state operating point calculations and added as a multi-dimensional look-up table to the fundamental flux model. The model extension can also be directly included in the IRTF model, shown in Fig. 3.7 and defined in (3.11a) - (3.11c). Thus, in both cases, the rotor current gets additional harmonics, damping its origin and influencing the NVH behavior of the induction machine.

$$\vec{\Psi}_{\mathrm{S}} = \sum_{|\nu|=1}^{\infty} \vec{\Psi}_{\mathrm{M}}^{\nu} + L_{\sigma,\mathrm{S}}\vec{i}_{\mathrm{s}} \tag{3.10a}$$

$$\vec{\Psi}_{\mathrm{R}}^{\nu} = \vec{\Psi}_{\mathrm{M}}^{\nu} - L_{\sigma,\mathrm{R}}^{\nu}\vec{i}_{\mathrm{R}}^{\nu} \tag{3.10b}$$

$$\frac{\mathrm{d}\vec{\Psi}_{\mathrm{R}}^{\nu}}{\mathrm{d}t} = R_{\mathrm{R}}^{\nu}\vec{i}_{\mathrm{R}}^{\nu} + j(1 - s_{\nu})\omega_{\mathrm{s}}\vec{\Psi}_{\mathrm{R}}^{\nu} \tag{3.10c}$$

$$\vec{\Psi}_{\mathrm{S}} = \sum_{|\nu|=1}^{\infty} \vec{\Psi}_{\mathrm{M}}^{\nu} + L_{\sigma,\mathrm{S}}\vec{i}_{\mathrm{s}} \tag{3.11a}$$

$$\vec{\Psi}_{\mathrm{R}}^{\mathrm{xy},\nu} = \vec{\Psi}_{\mathrm{M}}^{\mathrm{xy},\nu} - L_{\sigma,\mathrm{R}}^{\nu}\vec{i}_{\mathrm{R}}^{\mathrm{xy},\nu} \tag{3.11b}$$

$$\frac{\mathrm{d}\vec{\Psi}_{\mathrm{R}}^{\mathrm{xy},\nu}}{\mathrm{d}t} = R_{\mathrm{R}}^{\nu}\vec{i}_{\mathrm{R}}^{\mathrm{xy},\nu} \tag{3.11c}$$

Temporal rotor harmonics τ_{r} that are induced from spatial stator harmonics to the rotor current, due to the IRTF fundamental flux model with a rotor harmonics extension are defined with the stator frequency f_{s} as,

$$\tau_{\mathrm{r}} = \underbrace{(1 - \nu(1 - s))}_{s_{\nu}}f_{\mathrm{s}}. \tag{3.12}$$

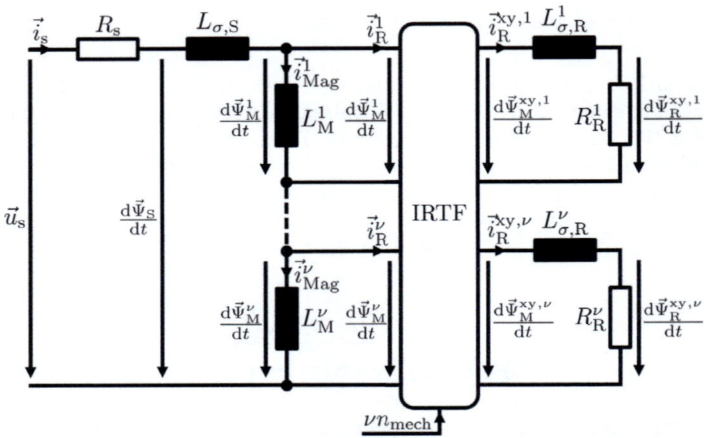

Figure 3.7: IRTF fundamental flux model with rotor harmonics extension

The amplitudes of the induced harmonics ν from the stator side with the temporal frequency of τ_r are shown in Fig. 3.8a for mean stator d-axis currents and in Fig. 3.8b for mean stator q-axis currents. Because of the neglected saturation harmonics, the amplitudes of the induced harmonics increase with the increasing fundamental rotor current, due to the increased induction of the stator harmonics ν.

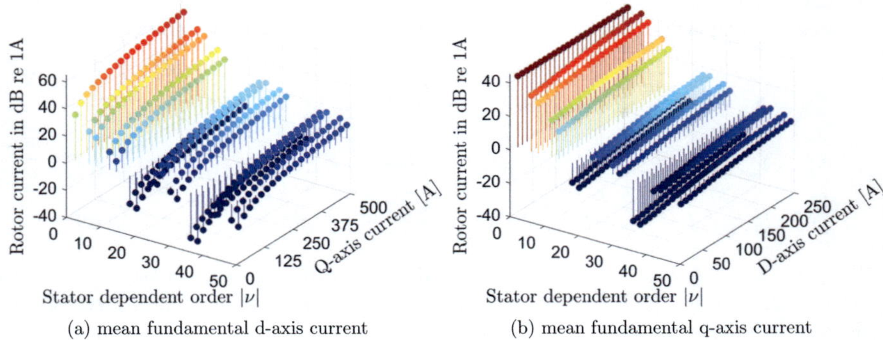

(a) mean fundamental d-axis current

(b) mean fundamental q-axis current

Figure 3.8: Stator dependent rotor current harmonics for mean fundamental (a) d-axis and (b) q-axis stator current

3.1.2 Force Model

After the calculation of the stator and rotor currents the stator mmf $\Theta_s(\alpha,t)$, the rotor mmf $\Theta_r(\alpha,t)$, the air-gap permeance $\Lambda_{air-gap}$ and finally the radial magnetic flux density $B_r(\alpha,t)$ are calculated [Bes08; Mal00; Le +09; GWC06; BHK15; Bis+16].

3.1.2.1 Stator Magneto Motive Force (mmf)

The basis for the stator mmf is the current of each stator conductor. Depending on the number of conductors in each slot, the spatial electric load $a_{el}(\alpha,t)$ distribution of the stator is calculated with the winding function $N_{el,s}^m(\alpha)w$. This function in turn, relates to the spatial winding scheme from Fig. 3.9, shown in (3.13), for $n_s = 6$ conductors in each slot, stator phase m and a current amplitude of 1A, shown in Fig. 3.10a. The spatial integration over the electric load results in the stator mmf, shown in (3.14) and Fig. 3.10b [MVP08; GWC06; BHK15; Bis+16; Ria].

Stator slot	1	2	3	4	5	6	7	8	9	10	11	12	13	14	15	16	17	18	19	20	21	22	23	24
Phase	V+	V+	V+	V+	U+	U+	U+	U−	W−	W−	W−	W−	V−	V−	V−	V−	U−	U−	U−	U−	W+	W+	W+	W+
	V+	V+	V+	V+	U+	U+	U+	U−	W−	W−	W−	W−	V−	V−	V−	V−	U−	U−	U−	U−	W+	W+	W+	W+

Figure 3.9: Stator winding scheme for one pole pair

$$a_{el,s}^m(\alpha,t) = n_s N_{el,s}^m(\alpha) i_s^m(t) \qquad (3.13)$$

$$\Theta_s(\alpha,t) = \sum_{m=1}^{3} \int_0^{2\pi} a_{el,s}^m(\alpha,t)\,d\alpha \qquad (3.14)$$

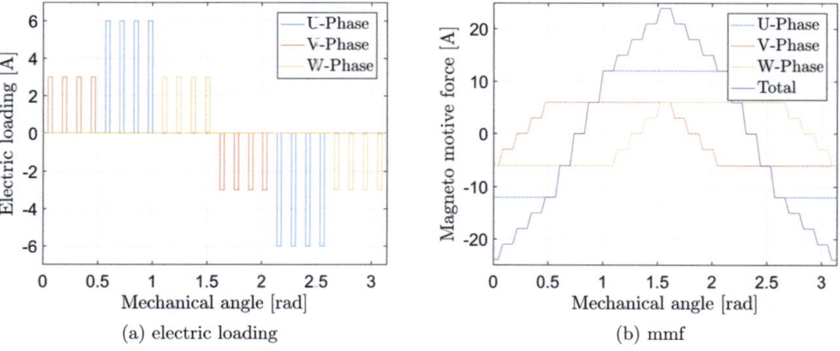

(a) electric loading

(b) mmf

Figure 3.10: Stator (a) electric loading and (b) mmf

The occurring spatial winding harmonics ν_{winding}, or pole pair relating $\tilde{\nu}_{\text{winding}}$, included in the stator mmf for the symmetrical star connected electrical machine with phase number m are [Lac05; GWC06; HRC05],

$$\boxed{\nu_{\text{winding}} = 1 \pm 2mk_\nu \text{ with } k_\nu \in \mathbb{Z}} \tag{3.15}$$

and

$$\tilde{\nu}_{\text{winding}} = p\nu_{\text{winding}}. \tag{3.16}$$

These equations are only concerning to the spatial winding harmonics. Spatial slotting harmonics, that are already included in these equations can be directly calculated with the number of stator slots Q_{s} as,

$$\nu_{\text{slot}} = 1 \pm \frac{Q_{\text{s}}k_\nu}{p} \text{ with } k_\nu \in \mathbb{Z} \tag{3.17}$$

and pole pair relating,

$$\tilde{\nu}_{\text{slot}} = p\nu_{\text{slot}}. \tag{3.18}$$

In order to include (3.14) into the simulation, spatial mmf functions $N_{\Theta,\text{s}}^{\text{m}}(\alpha)$ are created as for each phase m into,

$$N_{\Theta,\text{s}}^{\text{m}}(\alpha) = \int_0^{2\pi} a_{\text{el,s}}^{\text{m}}(\alpha,t)\bigg|_{i_{\text{s}}(t)=1\text{A}} \,\mathrm{d}\alpha. \tag{3.19}$$

The function can then, with the phase currents $i_{\text{s}}^{\text{m}}(t)$ and the total number of stator phases m_{s}, be inserted directly into,

$$\boxed{\Theta_{\text{s}}(\alpha,t) = \sum_{m=1}^{m_{\text{s}}} N_{\Theta,\text{s}}^{\text{m}}(\alpha)i_{\text{s}}^{\text{m}}(t), \ \Theta_{\text{s}}(\alpha,t) \in \mathbb{R}}. \tag{3.20}$$

Depending on the slot width b_{s} of the stator, the stator winding function for phase U is shown in Fig. 3.11. This function can also be created for the other two phases and extracted from look-up-tables during the simulation.

The stator mmf in the SFD can be calculated without the discretized functions, shown in (3.21). In the SFD the stator current vector $\vec{i}_{\text{s}}(t)$ and the current vector angle $\angle\vec{i}_{\text{s}}(t)$ are used instead of the three-phase currents. With these equations, the stator mmf can be directly calculated without generating look-up-tables with (3.19), using the winding factor $\xi_{\text{s},\nu}$ depending on the spatial stator harmonics ν and the total number of stator phases m_{s}, calculated with (3.15) [Bis+16].

$$\boxed{\underline{\Theta}_{\text{s}}(u,t) = \frac{2n_{\text{s}}m_{\text{s}}\xi_{\text{s},\nu}}{\pi\nu p}|\vec{\underline{i}}_{\text{s}}(t)|e^{\mathrm{j}\angle\vec{\underline{i}}_{\text{s}}(t)}, \ \underline{\Theta}_{\text{s}}(u,t) \in \mathbb{C}} \tag{3.21}$$

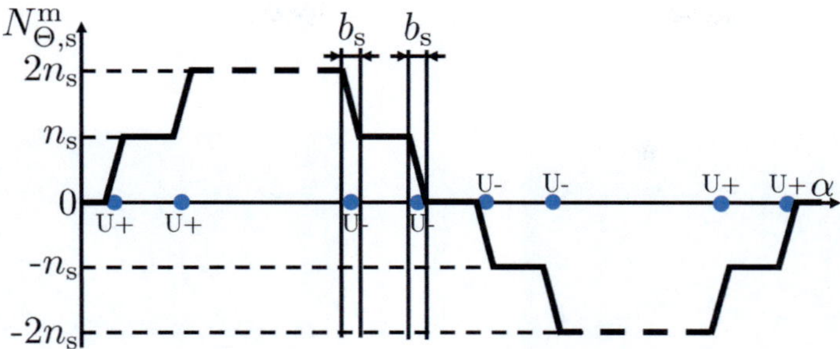

Figure 3.11: Stator winding function for U-Phase

The spatial comparison of the stator mmf calculation in spatial time domain (STD) from (3.20) and in spatial frequency domain (SFD) from (3.21) are shown in Fig. 3.12a and Fig. 3.12b for one time step. For the comparison in STD the simulation result from the SFD simulation is transformed back into the STD and vice versa. It is shown, that the same results like the discretized spatial time simulation can be achieved directly in the steady spatial frequency simulation. In this example, less than 100 spatial orders are considered.

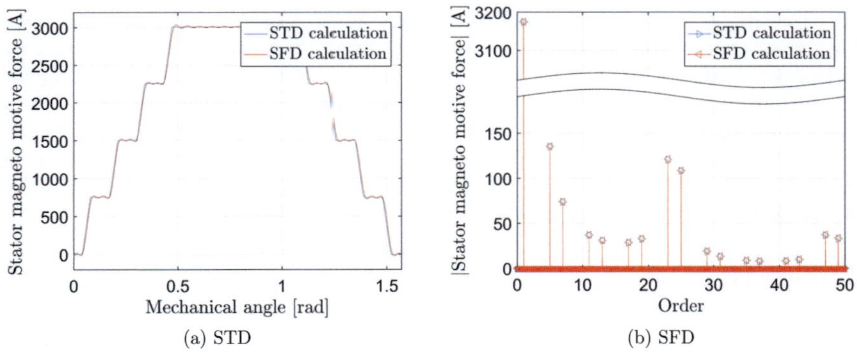

(a) STD

(b) SFD

Figure 3.12: Comparison of stator mmf calculations in (a) STD and (b) SFD

The temporal comparison of the spatial stator mmf simulations in STD and SFD are shown in Fig. 3.13a and 3.13b. There are small differences because of the discretization in the STD simulation. Due to the steady calculation in SFD, the FT based spectrum shows less calculation deviations. In this example, the stator is fed with sinusoidal stator currents, that causes a fundamental temporal order.

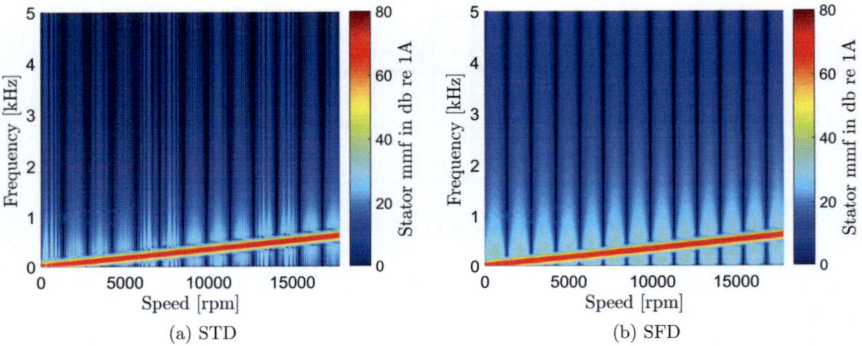

(a) STD   (b) SFD

Figure 3.13: Run-up of stator mmf simulations in (a) STD and (b) SFD

3.1.2.2 Rotor Magneto Motive Force (mmf)

The rotor magneto motive force (mmf) calculation bases on the rotor current of each rotor bar. Due to the squirrel-cage rotor, the number of rotor conductors is $n_r = 1$. With this information, the electric loading is calculated with (3.22), where q_r is the number of the relating rotor bar and $N_{el,r}^{q_r}(\alpha)$ is the spatial information relating to this rotor bar. Assuming a sinusoidal spatial rotor current distribution in (3.24), neglecting higher induced harmonics on the rotor bars, the electric loading can be calculated as shown in Fig. 3.14a. Equation (3.23) defines the rotor mmf calculation analogously to the stator mmf with integrating the electric loading of the rotor bar currents over the spatial angle, shown in Fig. 3.14b [JDP01; Bis+16; BHK15; HRC05; GWC06; Mal00; Bes08; GMA99; Bos+04].

$$a_{el,r}^{q_r}(\alpha,t) = n_r N_{el,r}^{q_r}(\alpha) i_r^{q_r}(t) \qquad (3.22) \qquad \Theta_r(\alpha,t) = \sum_{q_r=1}^{Q_r} \int_0^{2\pi} a_{el,r}^{q_r}(\alpha,t)\mathrm{d}\alpha \qquad (3.23)$$

$$i_r^{q_r}(t) = |\vec{i_r}| \cos\left(s\omega_s t - \frac{2p\pi}{Q_r} q_r\right) \qquad (3.24)$$

The occurring spatial rotor harmonics for a sinusoidal spatial distributed rotor mmf, neglecting the induction of the higher harmonics, are defined as,

$$\boxed{\mu_{\text{winding}} = \mu_{\text{slot}} = 1 \pm \frac{Q_r k_\mu}{p}, \text{ with } k_\mu \in \mathbb{Z}} \qquad (3.25)$$

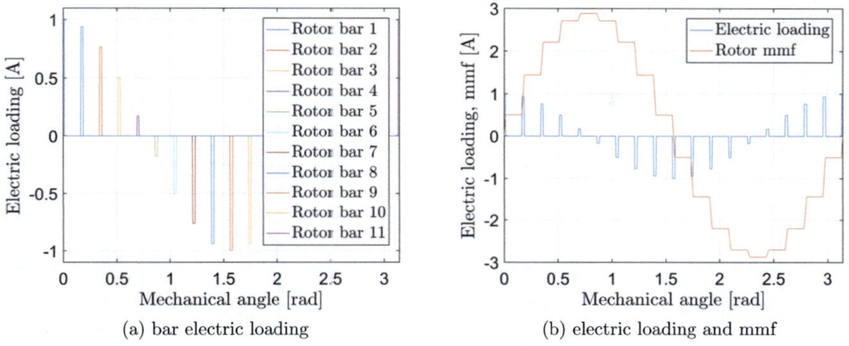

(a) bar electric loading

(b) electric loading and mmf

Figure 3.14: Rotor (a) bar electric loading and (b) electric loading, mmf

and pole pair relating as,

$$\tilde{\mu}_{\text{winding}} = \tilde{\mu}_{\text{slot}} = p\mu_{\text{winding}} = p\mu_{\text{slot}} \tag{3.26}$$

In order to save calculation time, the rotor mmf can be calculated using rotor winding functions $N_{\Theta,\text{r}}^{\text{qr}}(\alpha,t)$, that are moving with the mechanical rotor speed. The winding function can be implemented as two neighbored rotor bars, shown in Fig. 3.15a, and as two pole pair repeating rotor bars, shown in Fig. 3.15b, depending on the rotor slot width b_{qr}, based on (3.27).

$$N_{\Theta,\text{r}}^{\text{qr}}(\alpha,t) = \int_0^{2\pi} a_{\text{el,r}}^{\text{qr}}(\alpha,t)\bigg|_{i_\text{r}^{\text{qr}}(t)=1\text{A}} \, \mathrm{d}\alpha \tag{3.27}$$

The rotor mmf in STD, assuming a sinusoidal rotor current distribution, is defined as,

$$\boxed{\Theta_\text{r}(\alpha,t) = \sum_{\text{qr}=1}^{Q_\text{r}} N_{\Theta,\text{r}}^{\text{qr}}(\alpha,t)i_\text{r}^{\text{qr}}(t), \; \Theta_\text{r}(\alpha,t) \in \mathbb{R}.} \tag{3.28}$$

Analogous to the stator mmf calculation in SFD, the rotor mmf, depending on the rotor winding factor $\xi_{\text{r},\mu}$ and the mechanical rotor angle $\angle\theta_\text{r}(t)$, can be expressed by a sinusoidal rotor current distribution as a steady function in (3.29) [Bis+16].

$$\boxed{\underline{\Theta}_\text{r}(u,t) = \frac{Q_\text{r}\xi_{\text{r},\mu}}{2p\tau\mu}|\vec{i}_\text{r}(t)|e^{\text{j}(p\angle\theta_\text{r}(t)+\angle\vec{i}_\text{r}(t))}, \; \underline{\Theta}_\text{r}(u,t) \in \mathbb{C}} \tag{3.29}$$

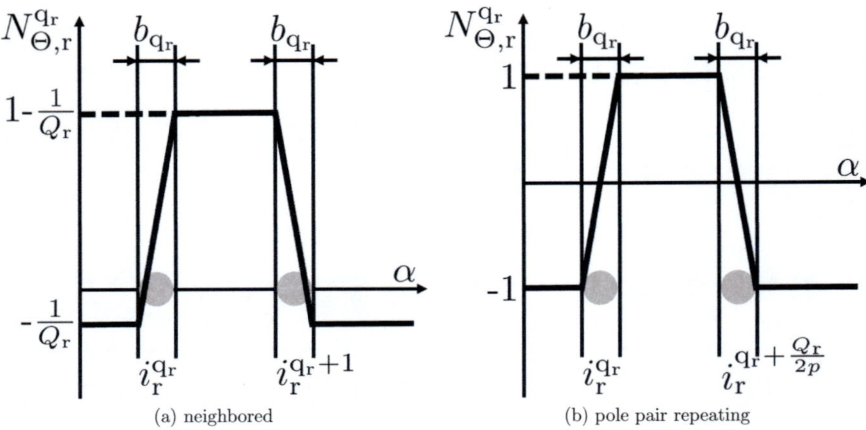

(a) neighbored

(b) pole pair repeating

Figure 3.15: Rotor winding function for two (a) neighbored and (b) pole pair repeating rotor bars

The temporal comparison of the rotor mmfs are shown in Fig. 3.16a in STD and in Fig. 3.16b in SFD. Similar to the temporal comparison of the stator mmfs, the rotor mmfs in STD, due to the discretization, shows small differences to the steady SFD simulation. Because of the finite number of 100 spatial rotor mmf orders in SFD, the number of visible temporal orders is limited. This does not occur in STD.

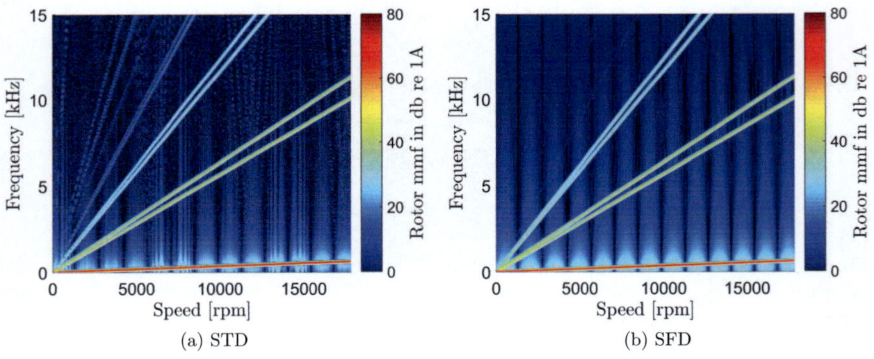

(a) STD

(b) SFD

Figure 3.16: Run-up of rotor mmf simulation in (a) STD and (b) SFD

Extension of the rotor mmf model

In order to include the extension of the fundamental flux model, from section 3.1.1.3, additional harmonics induced by the spatial stator harmonics are influencing the rotor mmf. The rotor current depending on these induced harmonics is defined as,

$$i_{\mathrm{r}}^{\mathrm{q_r}}(t) = \sum_{\nu=-\infty}^{\infty} |\vec{i}_{\mathrm{r}}^{\nu}| \cos\left(s_{\nu}\omega_{\mathrm{s}}t - \frac{2p\pi}{Q_{\mathrm{r}}}\mathrm{q_r}\right). \tag{3.30}$$

Thus, the rotor winding harmonics μ_{winding} have an additional dependency on ν, shown in (3.31) and pole pair relating in (3.32).

$$\boxed{\mu_{\mathrm{winding}} = \nu_{\mathrm{winding}} \pm \frac{Q_{\mathrm{r}}k_{\mu}}{p}, \text{ with } k_{\mu} \in \mathbb{Z}} \tag{3.31}$$

$$\tilde{\mu}_{\mathrm{winding}} = p\mu_{\mathrm{winding}} \tag{3.32}$$

The calculation in STD with the rotor current harmonics extension in (3.28) is still valid. For the calculation in SFD, the stator mmf equation has to consider these additional harmonics, as shown in (3.33) [Bis+16].

$$\boxed{\underline{\Theta}_{\mathrm{r}}(u,t) = \frac{Q_{\mathrm{r}}\xi_{\mathrm{r},\mu}}{2p\pi\mu}|\vec{i}_{\mathrm{r}}^{\nu}(t)|e^{\mathrm{j}(p\angle\theta_{\mathrm{r}}(t)+\angle\vec{i}_{\mathrm{r}}^{\nu}(t))}, \ \underline{\Theta}_{\mathrm{r}}(u,t) \in \mathbb{C}} \tag{3.33}$$

The spatial comparison between rotor mmf simulation in STD and SFD is shown in Fig. 3.17a and Fig. 3.17b for the extended model structure. For this investigation, the SFD calculation is transformed into STD and the STD calculation is transformed into SFD. The differences of the simulation results are occurring due to small winding factor differences. In the squirrel-cage rotor case, the fundamental winding factor $\xi_{\mathrm{r},1}$ is assumed to be 1. The exact calculation shows a fundamental winding factor for $\xi_{\mathrm{r},1} = 0.998$. This causes a small difference in the fundamental mmf amplitude.

In the STD, due to the induced higher harmonics into the rotor, the rotor mmf includes higher harmonics. These higher harmonics in the rotor current lead to a non-sinusoidal spatial current distribution. The higher current harmonics are also visible in the SFD. The orders of 5th, 7th, 11th, 13th,... would not be there with a sinusoidal rotor current, only the slotting harmonics of 17th, 19th,... would be visible, because of the $Q_{\mathrm{r}} = 36$ rotor bars.

3.1.2.3 Air-gap Permeance

The air-gap permeance $\Lambda_{\mathrm{air-gap}}$ can be seen as an inductance, the flux has to enter from the stator to the rotor side, or vice versa. Due to an accurate simulation, the air-gap

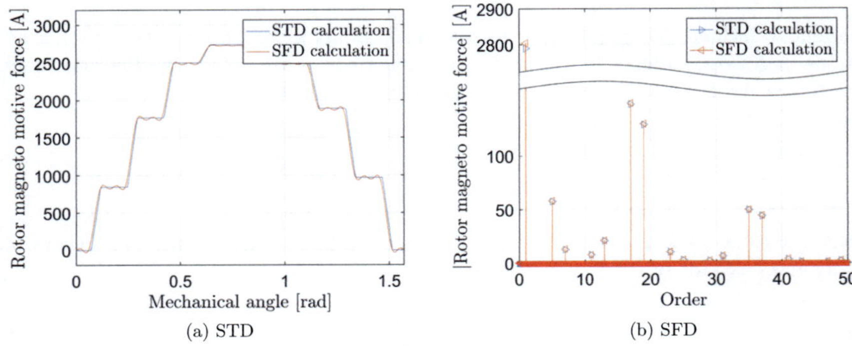

(a) STD

(b) SFD

Figure 3.17: Comparison of rotor mmf calculations in (a) STD and (b) SFD

permeance, with the vacuum permeability μ_0 and a fictive air-gap function g_{fic} is defined in STD as [Bes08; Mal00; BHK15; Bis+16; GWC06; BN09],

$$\boxed{\Lambda_{\text{air-gap}}(\alpha,t) = \frac{\mu_0}{g_{\text{fic}}(\alpha,t)}, \ \Lambda_{\text{air-gap}}(\alpha,t) \in \mathbb{R}.} \tag{3.34}$$

The fictive air-gap function $g_{\text{fic}}(\alpha,t)$ relates to the mechanical air-gap width δ, the stator and rotor slot function $f_{\text{s}}(\alpha)$ and $f_{\text{r}}(\alpha,t)$, as well as the factor of stator and rotor slot entering of the magnetic flux density e_{s} and e_{r}, that is defined in (3.35).

$$g_{\text{fic}}(\alpha,t) = \delta + e_{\text{s}}f_{\text{s}}(\alpha) + e_{\text{r}}f_{\text{r}}(\alpha,t) \tag{3.35}$$

The factor of the magnetic flux density entering the slots, depends on the stator and rotor slot width, b_{s} and b_{qr}. The slot width and subsequently the entering factor, can be set to zero for closed slotting, assuming deep buried stator conductors or rotor bars. In case of an open slotted stator and rotor, the factor of entering the slotting can be calculated as [Le +09],

$$e_{\text{s}} \approx \frac{b_{\text{s}}}{5} \tag{3.36a}$$

$$e_{\text{r}} \approx \frac{b_{\text{qr}}}{5} \tag{3.36b}$$

In this assumption, the magnetic flux density enters the stator and rotor slotting of a factor of 20%. This factor can be a good approximation, but in case of saturation, this factor has to be changed, in order to reach accurate results for every operating point of the induction machine.

The slotting functions of the stator and the rotor are defined similar to rectangular slotting. Thus, the slotting function is equal to 1 in case of a slot and 0 in case of a tooth, shown in Fig. 3.18a and Fig. 3.18b. The slotting functions for the stator and the rotor are therefore defined as,

$$f_\text{s}(\alpha) = f_\text{r}(\alpha) = \begin{cases} 1 & \text{if slot} \\ 0 & \text{if tooth} \end{cases}. \tag{3.37}$$

Moving the rotor slot function spatially with mechanical rotor speed n_mech relates to the temporal dependency t of the rotor slot function $f_\text{r}(\alpha,t)$.

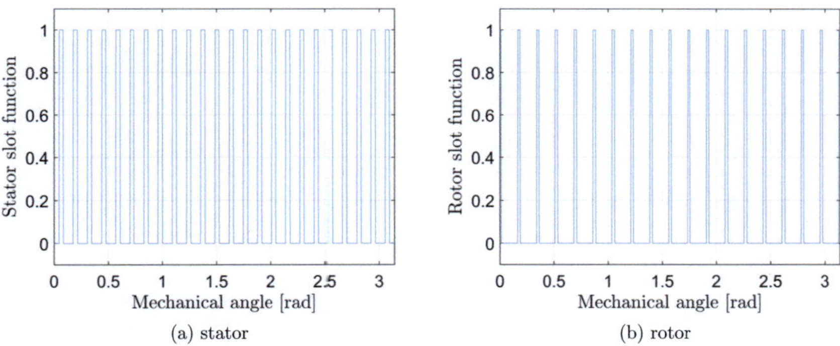

Figure 3.18: Spatial (a) stator and (b) rotor slot function

With this information, the spatial mechanical orders occurring due to the stator and rotor slotting, neglecting the interaction of each other, are defined as ν^* for the stator slots and μ^* for the rotor slots, as

$$\nu^* = \pm\frac{Q_\text{s}}{p}k_{\nu^*}, \text{ with } k_{\nu^*} \in \mathbb{Z} \tag{3.38a} \qquad \mu^* = \pm\frac{Q_\text{r}}{p}k_{\mu^*}, \text{ with } k_{\mu^*} \in \mathbb{Z} \tag{3.38b}$$

The calculation of the slotting function in SFD in case of closed stator or rotor slot is defined as,

$$\underline{f}_\text{s}(u) = \begin{cases} \frac{Q_\text{r}b_\text{s}}{2\pi} & \text{if } \nu^* = 0 \\ \frac{Q_\text{r}b_\text{s}}{\pi}\text{sinc}(\frac{\nu^*b_\text{s}}{\pi})e^{-\text{j}p\varphi} & \text{if } \nu^* \neq 0 \end{cases} \qquad \underline{f}_\text{r}(u,t) = \begin{cases} \frac{Q_\text{r}b_\text{qr}}{2\pi} & \text{if } \mu^* = 0 \\ \frac{Q_\text{r}b_\text{qr}}{\pi}\text{sinc}(\frac{\mu^*b_\text{s}}{\pi})e^{-\text{j}p\mu^*\angle\theta_\text{r}(t)} & \text{if } \mu^* \neq 0 \end{cases}$$

$$\tag{3.39a} \qquad\qquad\qquad\qquad \tag{3.39b}$$

The resulting air-gap permeance in SFD for closed rotor slots is shown in (3.40a) and for closed stator slots in (3.40b). For example the air-gap permeance for closed rotor slots is shown in Fig. 3.19a in STD and in Fig. 3.19b in SFD. It is shown, that the air-gap permeance calculated directly in spatial frequency domain fits to the calculation in spatial time domain.

$$\underline{\Lambda}_{\text{air}-\text{gap}}(u) = \begin{cases} \frac{\mu_0}{\delta}\left(1 - \left(1 - \frac{1}{1+\frac{e_s}{\delta}}\right)\underline{f}_s(u)\right) & \text{if } \nu^* = 0 \\ \frac{\mu_0}{\delta}\left(1 - \frac{1}{1+\frac{e_s}{\delta}}\right)\underline{f}_s(u) & \text{if } \nu^* \neq 0 \end{cases} \tag{3.40a}$$

$$\underline{\Lambda}_{\text{air}-\text{gap}}(u,t) = \begin{cases} \frac{\mu_0}{\delta}\left(1 - \left(1 - \frac{1}{1+\frac{e_r}{\delta}}\right)\underline{f}_r(u,t)\right) & \text{if } \mu^* = 0 \\ \frac{\mu_0}{\delta}\left(1 - \frac{1}{1+\frac{e_r}{\delta}}\right)\underline{f}_r(u,t) & \text{if } \mu^* \neq 0 \end{cases} \tag{3.40b}$$

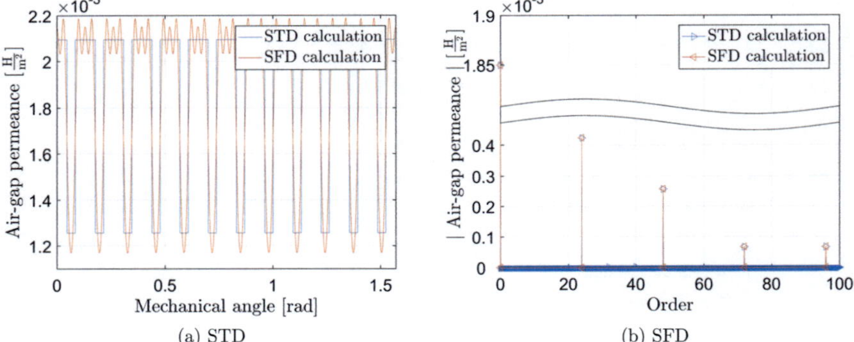

(a) STD

(b) SFD

Figure 3.19: Air-gap permeance for closed rotor slots in (a) STD and (b) SFD

As it is shown in Fig. 3.20a, the magnetic permeance can be calculated considering a few spatial orders. Fig. 3.20b shows the spatial decomposition of the air-gap permeance. It can be seen, with the fact of both stator and rotor opened slots, there are harmonics because of the interaction of the stator and rotor slots. Amplitudes, associated to either stator or rotor side effects, are influenced as well. Because of unrepresentable analytical expressions for induction machines, with both stator and rotor opened slots, the Fourier transform into SFD can be solved in a pre-calculation for every rotor position, as shown in 3.41.

$$\underline{\Lambda}_{\text{air}-\text{gap}}(u,t) = \frac{1}{2\pi}\int_0^{2\pi}\frac{\mu_0}{g_{\text{fic}}(\alpha,t)}e^{-ju\alpha}\mathrm{d}\alpha, \quad \underline{\Lambda}_{\text{air}-\text{gap}}(u,t) \in \mathbb{C} \tag{3.41}$$

With the fact of interaction, the magnetic air-gap permeance includes additionally the mechanical interaction harmonics of stator and rotor slots, defined as

$$\nu_{\text{int}}^* = \nu^* + \mu^*. \tag{3.42}$$

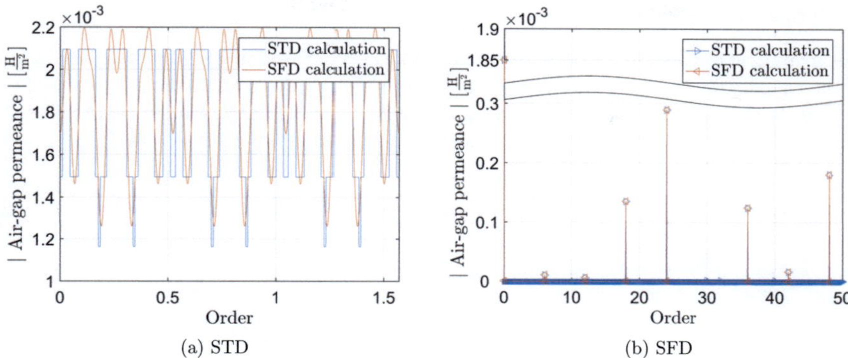

Figure 3.20: Air-gap permeance for stator and rotor opened slots in (a) STD and (b) SFD

3.1.2.4 Saturation Calculation

In order to include the saturation behavior and with it the saturation harmonics of the induction machine in the force model, the saturation has to be calculated additionally. It is shown how to calculate the saturation based on analytical equations, electromagnetic FEM simulations, as well as the saturation based on the material curves [Mal00; Bes08; GWC06; MN83; Lee61; ML92; Cha+96; LT74; Don+99].

The saturation $\Lambda_{\mathrm{sat}}(\alpha,t)$ is added to the magnetic air-gap permeance $\Lambda_{\mathrm{air-gap}}(\alpha,t)$ in STD, as defined in

$$\Lambda_{\mathrm{air-gap}}(\alpha,t) = \frac{\mu_0}{g_{\mathrm{fic}}(\alpha,t)} + \Lambda_{\mathrm{sat}}(\alpha,t),\ \Lambda_{\mathrm{air-gap}}(\alpha,t) \in \mathbb{R}. \tag{3.43}$$

and as saturation $\Lambda_{\mathrm{sat}}(u,t)$ in SFD, as

$$\underline{\Lambda}_{\mathrm{air-gap}}(u,t) = \frac{1}{2\pi} \int_0^{2\pi} \frac{\mu_0}{g_{\mathrm{fic}}(\alpha,t)} e^{-\mathrm{j}u\alpha}\,\mathrm{d}\alpha + \underline{\Lambda}_{\mathrm{sat}}(u,t),\ \underline{\Lambda}_{\mathrm{air-gap}}(u,t) \in \mathbb{C} \tag{3.44}$$

With this information, the saturation harmonics are convoluted with the spatial harmonics from stator and rotor mmfs, ν and μ. The occurring spatial harmonics are defined as,

$$\nu' = \nu + 2\nu_{\mathrm{sat}},\ \text{with } \nu_{\mathrm{sat}} \in \mathbb{Z} \tag{3.45a} \qquad \mu' = \mu + 2\mu_{\mathrm{sat}},\ \text{with } \mu_{\mathrm{sat}} \in \mathbb{Z} \tag{3.45b}$$

Analytical saturation expression

The analytical expression for the saturation is based on the local magneto motive force parts of the air-gap Θ_g, the rotor tooth Θ_{Rz}, the stator tooth Θ_{Sz}, as well as the stator and rotor yoke Θ_{Sj} and Θ_{Rj} over one pole division τ_p, shown in Fig. 3.21 [Sei92; Mal00; Bes08; Bis+16; ML92]. The calculation of saturation $\Lambda_{sat}(\alpha,t)$ and $\Lambda_{sat}(u,t)$ are expressed as,

$$\Lambda_{sat}(\alpha,t) = \sum_{n=-\infty}^{\infty} \frac{\mu_0 k_{sat,n}(t)}{\delta} \sin(n\omega_s t + np\alpha + \varphi_{sat}(t)), \text{ where } \{2n : n \in \mathbb{Z}\} \tag{3.46a}$$

$$\underline{\Lambda}_{sat}(u,t) = \frac{\mu_0 k_{sat,u}(t)}{\delta} e^{-ju\omega_s + \varphi_{sat}(t)}, \text{ where } \{2u : u \in \mathbb{Z}\} \tag{3.46b}$$

The saturation coefficients for the main saturation waves $k_{sat,0}(t)$, as well as $k_{sat,2}(t)$ are defined as,

$$k_{sat,0}(t) = \frac{\Theta_g(t)}{\Theta_g(t) + 2\Theta_{Sz}(t) + 2\Theta_{Rz}(t) + \Theta_{Sj}(t) + \Theta_{Rj}(t)} \tag{3.47a}$$

$$k_{sat,2}(t) = k_{sat,0}(t)\frac{2A_{sat}(t)}{1 + A_{sat}(t)}, \text{ with } A_{sat}(t) = \frac{\Theta_{Sz}(t) + \Theta_{Rz}(t)}{\Theta_g(t) + \Theta_{Sj}(t) + \Theta_{Rj}(t)} \tag{3.47b}$$

In order to picture the analytical saturation behavior, an example for a normalized magnetic flux density is given in Fig. 3.22. The saturation of 0th order reduces the magnetic flux density in general, where the saturation of 2nd order adds a 3rd temporal and spatial order to the system.

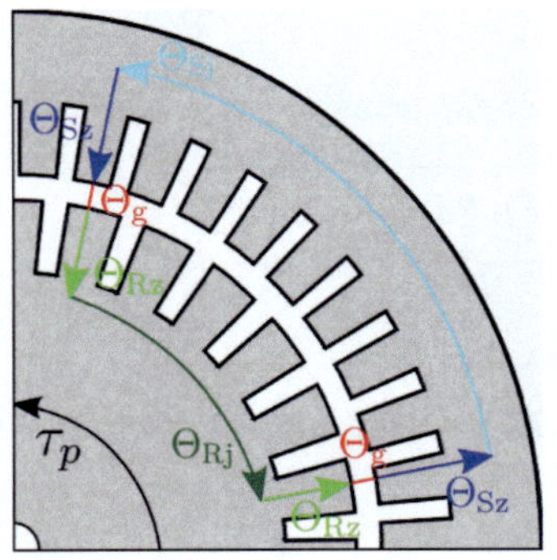

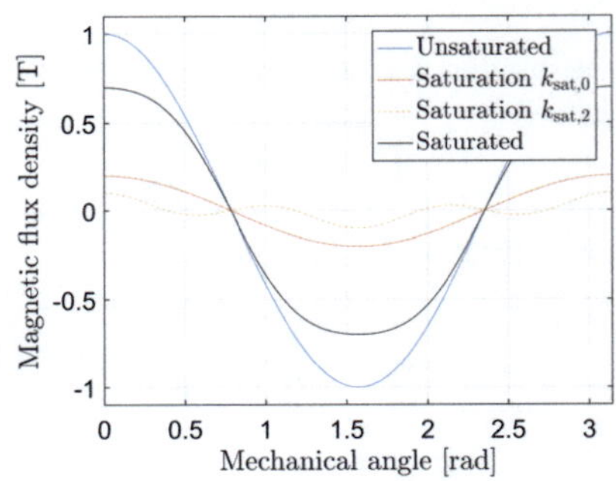

Figure 3.21: Magneto motive force parts for saturation calculation

Figure 3.22: Schematic of the analytical saturation simulation

The saturated air-gap permeance is shown in STD in Fig. 3.23a and in SFD in Fig. 3.23b. It is shown, that the spatial orders are completely matching. The comparison in STD shows a small deviation of the position in spatial time and spatial frequency domain. This deviation is caused by the angle φ_{sat}, which is on the one hand calculated with the phase shift of $\varphi_{\mathrm{sat}}(t) = \angle(\Theta_{\mathrm{s}}(\alpha,t) - \Theta_{\mathrm{r}}(\alpha,t))$ and on the other hand with $\varphi_{\mathrm{sat}}(t) = \angle(\underline{\Theta}_{\mathrm{s}}(1,t) - \underline{\Theta}_{\mathrm{r}}(1,t))$. In this case, the calculation in the SFD shows more accurate results, because of the fact, that the saturation should occur on position of the fundamental of the summed mmf $\Theta(\alpha,t) = (\Theta_{\mathrm{s}}(\alpha,t) - \Theta_{\mathrm{r}}(\alpha,t))$. In SFD this position is directly clarified by the use of the complex amplitudes of the spatial orders.

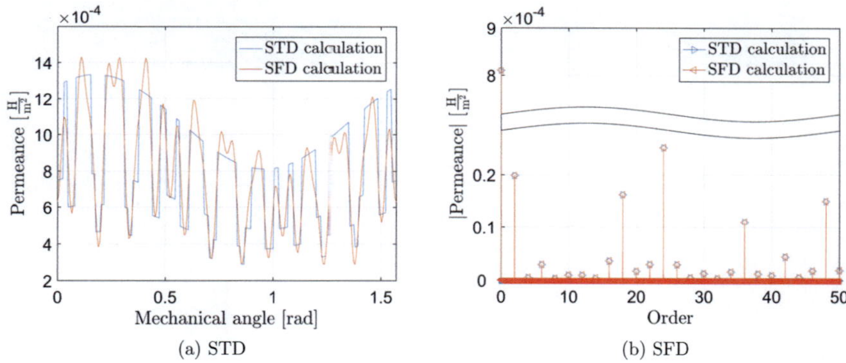

(a) STD   (b) SFD

Figure 3.23: Saturated magnetic permeance in (a) STD and (b) SFD

The temporal comparison of the magnetic permeance in STD and SFD, in Fig. 3.24a and 3.24b, show the validity of the calculation methods, as well as the reduction of discretization errors. The SFD is limited to 100 spatial orders, whereas the STD consists of infinite number of spatial orders.

A more accurate alternative to calculate the saturation coefficients is, to extract these coefficients from electromagnetic FEM simulations. For this calculation method, the FEM calculation is completed with air in the rotor slots to receive a proportionality of the stator current d-axis system relating to the magnetic flux density $i_{\mathrm{d}}(t) \propto \Psi(t) \propto B(\alpha,t)$. Using the spatial stator current distribution of the FEM and calculating the magnetic flux density with,

$$B_{\mathrm{r}}(\alpha,t) = \Theta(\alpha,t)\frac{\mu_0}{\delta}, \text{ where } \Theta_{\mathrm{r}}(\alpha,t) = 0 \text{ and } \Theta(\alpha,t) = \Theta_{\mathrm{s}}(\alpha,t). \tag{3.48}$$

The saturation coefficients can be extracted from the comparison of the calculated spatial magnetic flux density and the spatial magnetic flux density, created with (3.48), shown in Fig. 3.25 and 3.26. The coefficient $k_{\mathrm{sat},0}(t)$ and $k_{\mathrm{sat},2}(t)$ are extracted assuming that the constant part of saturation $k_{\mathrm{sat},0}(t)$ is mostly occurring at the outer ends of one pole and the coefficient $k_{\mathrm{sat},2}(t)$ of saturation is occurring in the middle of one pole after the reduction of the constant part of saturation.

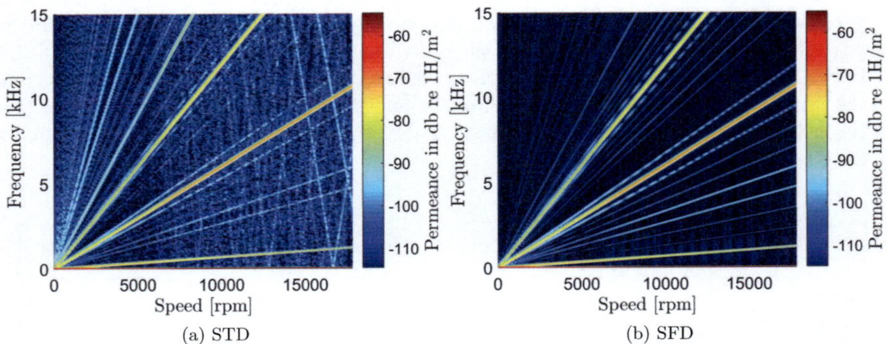

(a) STD

(b) SFD

Figure 3.24: Run-up of the magnetic permeance for (a) STD and (b) SFD calculation

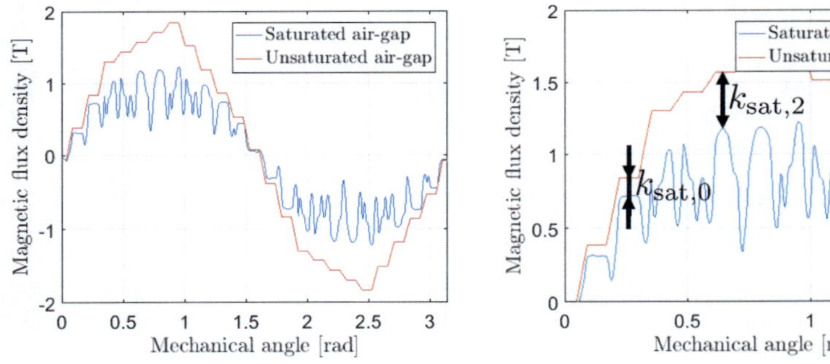

Figure 3.25: FEM comparison of unsaturated constant air-gap flux density and saturated flux density

Figure 3.26: Coefficient estimation based on FEM magnetic flux density simulation

The saturation can also be extracted using the temporal time behavior of the magnetic flux densities directly from the FEM calculation and the calculation from 3.48 beneath one stator tooth. This estimation is based on the assumption, that the spatial amplitude of the saturation is leading to the same behavior in the temporal time domain (TTD) and the saturation behavior beneath one tooth approximates the main behavior. The magnetic flux densities can be extracted by comparing in the temporal frequency domain (TFD). The disadvantage of this procedure is the convolution of the temporal harmonics. For instance, the slotting harmonics with the saturation harmonics change the amplitudes of the saturation coefficients. Nevertheless, all kind of analytical calculations are more or less approximations and can be better expressed with electromagnetic FEM simulations.

Material curve based saturation simulation

Another method in order to calculate the saturation behavior of the induction machine, is to spatially compare the mmf $\Theta(\alpha,t)$ in the online simulation to a material curve for each time step. This will lead to more simulation time, depending on the number of discretized spatial points, shown in Fig. 3.27 and 3.28. At the same time, it increases the accuracy of the temporal and spatial dependent simulation [BN09; Bis+17].

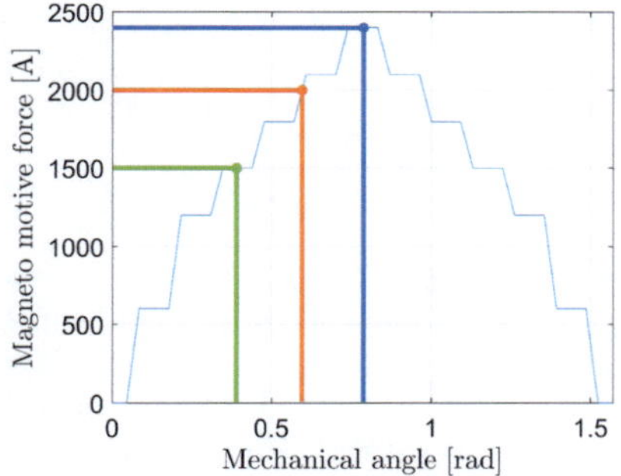

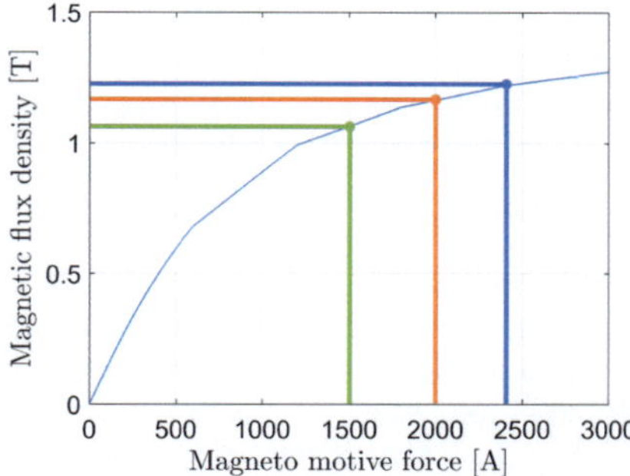

Figure 3.27: Comparison of mmf to material curve for saturation simulation

Figure 3.28: Material curve compared to mmf for saturation simulation

With this information, a non-linear curve saturation factor $f_{\text{sat}}(\Theta(\alpha,t))$ can be extracted from the material curve, in order to create a non-linear $\Lambda_{\text{air}-\text{gap}}(\alpha,t)$, as

$$\Lambda_{\text{air}-\text{gap}}(\alpha,t) = \frac{\mu_0}{\mathcal{g}_{\text{fic}}(\alpha,t)} f_{\text{sat}}(\Theta(\alpha,t)), \ \Lambda_{\text{air}-\text{gap}}(\alpha,t) \in \mathbb{R}. \tag{3.49}$$

The $B - \Theta$ material curve can be created using a $B - H$ material curve, where $\Theta \propto H$ and scaling it because of the flux path length through the machine. On the one hand, this procedure has the highest calculation time. On the other hand, it is the most accurate method in order to include saturation in this kind of simulation method. This procedure is not performable in the spatial frequency domain, due to the comparison of the spatial discretized points in the air-gap to the $B - H$ or $B - \Theta$ material curve in the STD.

3.1.2.5 Eccentricity Calculation

The eccentricities can be included changing the magnetic air-gap permeance $\Lambda_{\text{air}-\text{gap}}(\alpha,t)$. There are two different types of eccentricities, that change the electromagnetic behavior of the induction machine. The first type of eccentricity is the static eccentricity. It occurs

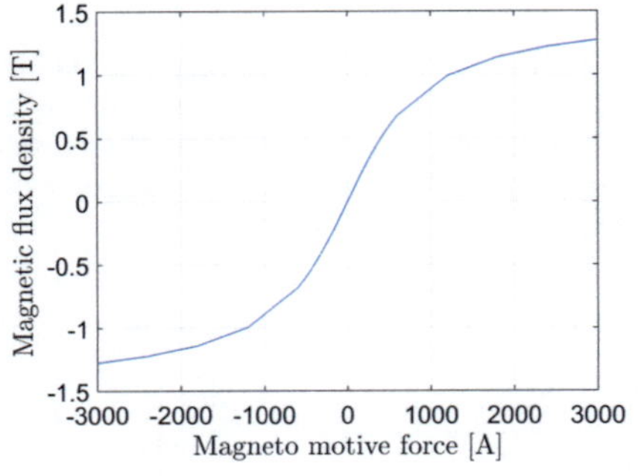

Figure 3.29: $B - \Theta$ material curve

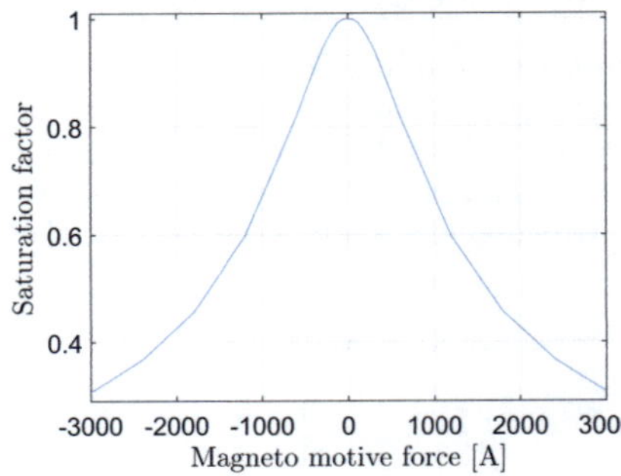

Figure 3.30: Non linear saturation factor

due to a constant eccentric position of the rotor or due to deformations of the stator. The second type of eccentricity is the dynamic eccentricity, which occurs due to an axial slanted rotor position, unbalanced rotor or one-sided magnetic pull from the static eccentricity. This calculation procedure is based on the assumption of the air-gap width $\delta \ll$ inner stator radius $r_{\mathrm{s,inner}}$. The schematic of a eccentric rotor position is shown in Fig. 3.31 and 3.32, where an uniform and healthy normalized magnetic flux density gets a additional harmonic due to the non-uniformity of the air-gap [GWC06; Bis+16; PLF12b; PEE15].

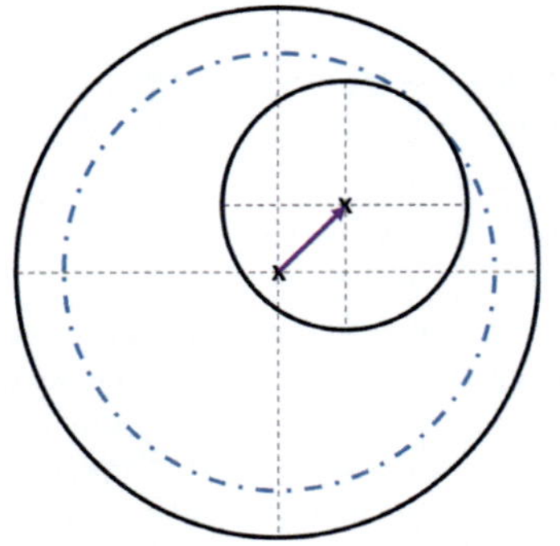

Figure 3.31: Schematic of eccentric rotor position

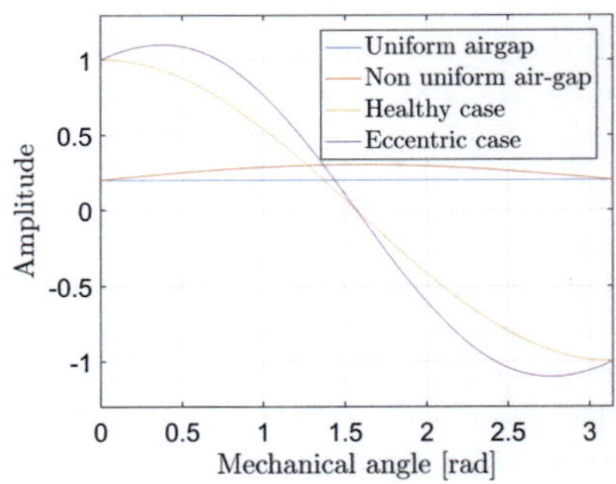

Figure 3.32: Flux behavior for eccentric rotor position

Static eccentricity

The static eccentricity leads to a spatially constant inequality of the air-gap width. This behavior can be included in the air-gap permeance $\Lambda_{\mathrm{air-gap}}(\alpha,t)$, with the static eccentric radius r_{stat}, as

$$\boxed{\Lambda_{\mathrm{air-gap}}(\alpha,t) = \frac{\mu_0}{g_{\mathrm{fic}}(\alpha,t)(1 - r_{\mathrm{stat}}\cos(\alpha))}, \quad \Lambda_{\mathrm{air-gap}}(\alpha,t) \in \mathbb{R}.}$$
(3.50)

and leads to a convolution of an additional order in the SFD as

$$\boxed{\underline{\Lambda}_{\mathrm{air-gap}}(u,t) = \frac{1}{2\pi}\int_0^{2\pi} \frac{\mu_0}{g_{\mathrm{fic}}(\alpha,t)}\epsilon^{-\mathrm{j}u\alpha}\mathrm{d}\alpha * (1 - \frac{\mu_0}{\delta}r_{\mathrm{stat}})e^{-\mathrm{j}\varphi_{\mathrm{stat}}}, \quad \underline{\Lambda}_{\mathrm{air-gap}}(u,t) \in \mathbb{C}.}$$
(3.51)

Spatial occurring orders, due to the static eccentricity are added to the mechanical stator and rotor slotting order, ν^* and μ^*, as

$$\nu^* = \nu^* \pm \frac{1}{p} \qquad (3.52\mathrm{a}) \qquad\qquad \mu^* = \mu^* \pm \frac{1}{p} \qquad (3.52\mathrm{b})$$

Fig. 3.33a and 3.33b show the magnetic permeance in case of a static eccentric rotor position. It can be regarded, that due to the convolution, additional harmonics are added to the system.

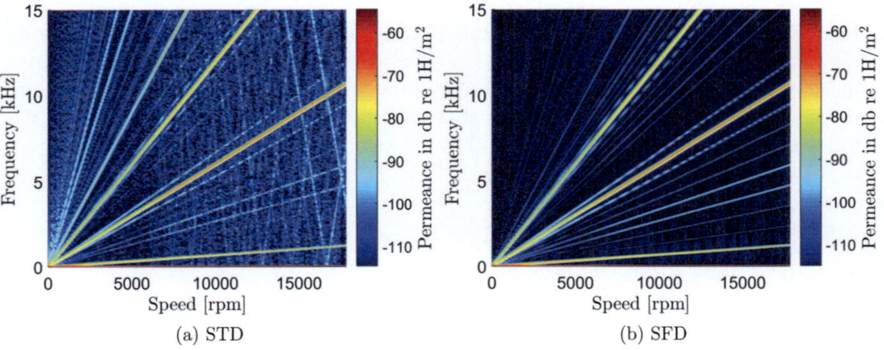

(a) STD   (b) SFD

Figure 3.33: Static eccentric air-gap permeance in a) STD and b) SFD

Dynamic eccentricity

The dynamic eccentricity changes the air-gap width depending on the rotor position. With this fact, the rotor inequality changes with the dynamic eccentric radius r_{dyn} depending on each time step. This behavior is described in the STD in (3.53).

$$\boxed{\Lambda_{\mathrm{air-gap}}(\alpha,t) = \frac{\mu_0}{g_{\mathrm{fic}}(\alpha,t)(1 - r_{\mathrm{dyn}}\cos(\alpha - \omega_{\mathrm{m}}t))}, \quad \Lambda_{\mathrm{air-gap}}(\alpha,t) \in \mathbb{R}}$$
(3.53)

It can also be expressed in the steady SFD with (3.54).

$$\underline{\Lambda}_{\text{air}-\text{gap}}(u,t) = \frac{1}{2\pi} \int_0^{2\pi} \frac{\mu_0}{g_{\text{fic}}(\alpha,t)} e^{-\text{ju}\alpha} \mathrm{d}\alpha * (1 - \frac{\mu_0}{\delta} r_{\text{dyn}}) e^{-\text{j}(\omega_{\text{m}}t+\varphi_{\text{dyn}})}, \ \underline{\Lambda}_{\text{air}-\text{gap}}(u,t) \in \mathbb{C}$$

(3.54)

The occurring spatial orders are the same as shown in (3.52a) and (3.52b). In this case of eccentricity, there are temporal harmonics occurring as well. The figures for one time step would seem to be the same like in the static eccentricity case, but the minima and maxima are moving with the mechanical rotor speed.

3.1.2.6 Skewing Calculation

Skewing can reduce the harmonic excitation of the electromagnetic flux density, which can subsequently reduce the induced harmonics into the stator and rotor coils, torque harmonics, as well as harmonics in the electromagnetic force excitation. Depending on the machine type, skewing can be performed for the stator or rotor side [GWC06; TA01; MP09; Kaw+09; Des+16].

In case of a force model, considering the 2-D effects and generalizing the local effects of skewing, the simulation can directly be performed in SFD. There the skewing is directly included in the winding factor of stator $\xi_{\text{s},\nu}$, in the rotor winding factor $\xi_{\text{r},\nu}$ for the mmf, in the magnetic permeance as an additional skewing factor of the mechanical rotor. In order to include the local effects of skewing, the most common used model structure is the axial segmentation of the electrical machine.

Segmentation of the induction machine

With the method of the axial segmentation of the induction machine, the skewing of stator and rotor slotting, as well as the harmonics excitation of the mmfs can be implemented in the force model. Considering a axial skewing angle $\varphi_{\text{skw}} = 10°$, the axial length l_{Fe} of the machine can be divided into $n_{\text{skw}} = 5$ parts, shown in Fig. 3.34. The skew part k_{skw} of the air-gap permeance segment $\Lambda_{\text{air}-\text{gap},k_{\text{skw}}}(\alpha,t)$ in spatial time domain can be defined as shown in (3.55). In addition, it can be generalized for the whole axial length in (3.56). The calculation in

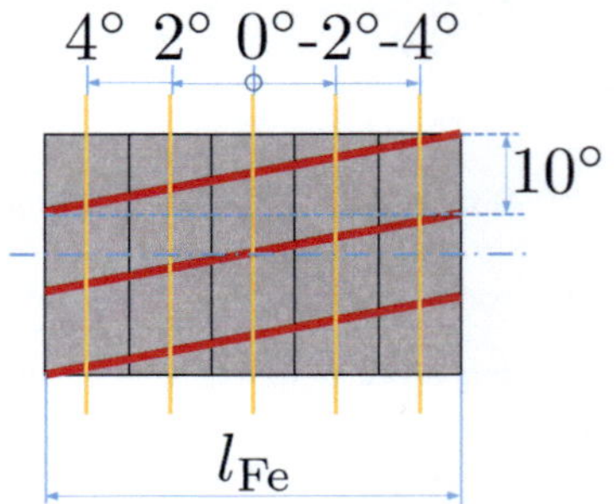

Figure 3.34: Schematic of axial segmentation

SFD can also be conducted with (3.41) adding the skewing angle φ_{skw} to the equation [GWC06; MP09; Ede+07].

$$\Lambda_{\mathrm{air-gap},\mathrm{k_{skw}}}(\alpha,t) = \frac{\mu_0}{g_{\mathrm{fic}}(\alpha - \alpha_{\mathrm{skw}}(\frac{k_{\mathrm{skw}}-1}{n_{\mathrm{skw}}-1}l_{\mathrm{Fe}} - \frac{l_{\mathrm{Fe}}}{2}),t)}, \ \Lambda_{\mathrm{air-gap},\mathrm{k_{skw}}}(\alpha,t) \in \mathbb{R} \tag{3.55}$$

$$\Lambda_{\mathrm{air-gap}}(\alpha,t) = \frac{1}{n_{\mathrm{skw}}} \sum_{k_{\mathrm{skw}}=1}^{n_{\mathrm{skw}}} \Lambda_{\mathrm{air-gap},\mathrm{k_{skw}}}(\alpha - \alpha_{\mathrm{skw}}(\frac{k_{\mathrm{skw}}-1}{n_{\mathrm{skw}}-1}l_{\mathrm{Fe}} - \frac{l_{\mathrm{Fe}}}{2}),t) \tag{3.56}$$

3.1.2.7 Magnetic Flux Density

After the calculation of the stator and rotor mmf, $\Theta_{\mathrm{s}}(\alpha,t)$ and $\Theta_{\mathrm{r}}(\alpha,t)$, as well as the magnetic air-gap permeance $\Lambda_{\mathrm{air-gap}}(\alpha,t)$, the radial magnetic flux density in the air-gap $B_{\mathrm{air-gap}}(\alpha,t)$ of the induction machine is defined in STD as [Sei92; MP09; GWC06; Bes08; BHK15],

$$B_{\mathrm{air-gap}}(\alpha,t) = \underbrace{(\Theta_{\mathrm{s}}(\alpha,t) + \Theta_{\mathrm{r}}(\alpha,t))}_{\Theta(\alpha,t)}\Lambda_{\mathrm{air-gap}}(\alpha,t), \ B_{\mathrm{air-gap}}(\alpha,t) \in \mathbb{R}. \tag{3.57}$$

In the spatial frequency domain the calculation of the magnetic flux density $\underline{B}_{\mathrm{air-gap}}(u,t)$ consists of a convolution of the summed up mmf $\underline{\Theta}(u,t)$ and the air-gap permeance $\underline{\Lambda}_{\mathrm{air-gap}}(u,t)$, shown in (3.58) [Bis+16].

$$\underline{B}_{\mathrm{air-gap}}(u,t) = \underbrace{(\underline{\Theta}_{\mathrm{s}}(u,t) + \underline{\Theta}_{\mathrm{r}}(u,t))}_{\underline{\Theta}(u,t)} * \underline{\Lambda}_{\mathrm{air-gap}}(u,t), \ \underline{B}_{\mathrm{u}}(u,t) \in \mathbb{C}. \tag{3.58}$$

The comparison of the STD and SFD model is shown in Fig. 3.35a and 3.35b. Deviations occur due to the limitation of the STD model of 100 spatial orders and due to the deviation of the saturation of the magnetic air-gap permeance.

The same behavior can be seen in the temporal comparison of the two domains. There are deviations of the spatial order limitation. However, the occurring error due to the discretization in the spatial time domain is minimized in the steady spatial frequency domain.

Flux linkage calculation

With the calculation of the electromagnetic flux density $B_{\mathrm{air-gap}}(\alpha,t)$, the flux linkages linking to the m-th stator coil $\psi_{\mathrm{m}}^{\mathrm{m}}(t)$ in the stator frame system with the linked radius r [Bis+17; MP09; Wol+15], defined as

$$\psi_{\mathrm{m}}^{\mathrm{m}}(t) = rl_{\mathrm{Fe}} \int_{0}^{2\pi} N_{\Theta,\mathrm{s}}^{\mathrm{m}}(\alpha) B_{\mathrm{air-gap}}(\alpha,t)\mathrm{d}\alpha, \ \psi_{\mathrm{m}}^{\mathrm{m}}(t) \in \mathbb{R}. \tag{3.59a}$$

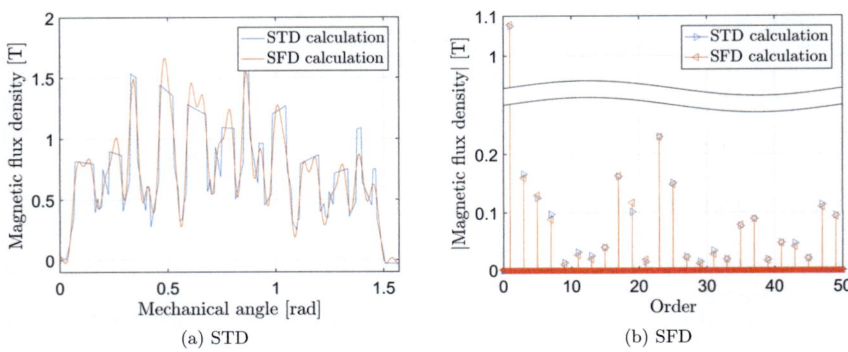

Figure 3.35: Spatial magnetic flux density in (a) STD and (b) SFD

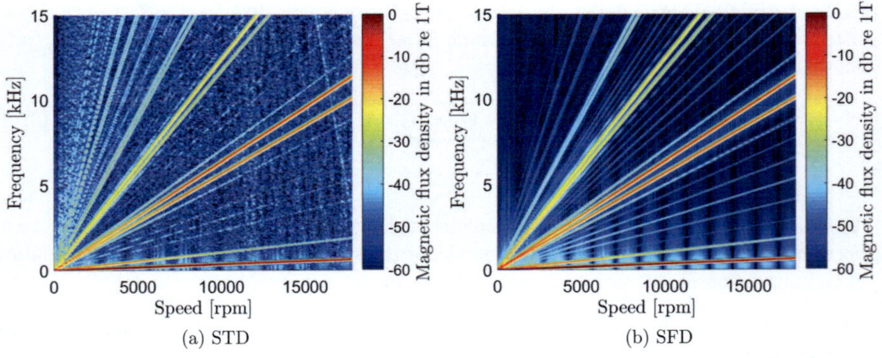

Figure 3.36: Run-up of the magnetic flux density in (a) STD and (b) SFD

The flux linkages linking to q_r-th rotor bars $\psi_{\mathrm{m}}^{\mathrm{q_r,xy}}(t)$ in the rotor frame system are defined as,

$$\psi_{\mathrm{m}}^{\mathrm{q_r,xy}}(t) = r l_{\mathrm{Fe}} \int_0^{2\pi} N_{\Theta,\mathrm{r}}^{\mathrm{q_r}}(\alpha,t) B_{\mathrm{air-gap}}(\alpha,t)\mathrm{d}\alpha, \ \psi_{\mathrm{m}}^{\mathrm{q_r,xy}}(t) \in \mathbb{R}. \tag{3.60a}$$

In the spatial frequency domain the stator and rotor flux linkages can be directly defined as shown in (3.61) and (3.62).

$$\underline{\psi}_{\mathrm{m}}^{\mathrm{m}}(t) = rl_{\mathrm{Fe}} \sum_{u=-\infty}^{\infty} N_{\Theta,\mathrm{s}}^{\mathrm{m}}(u)\underline{B}_{\mathrm{air-gap}}(u,t)$$

(3.61)

$$\underline{\psi}_{\mathrm{m}}^{\mathrm{qr,xy}}(t) = rl_{\mathrm{Fe}} \sum_{u=-\infty}^{\infty} \underline{N}_{\Theta,\mathrm{r}}^{\mathrm{qr}}(u,t)\underline{B}_{\mathrm{air-gap}}(u,t)$$

(3.62)

Within the SFD, $N_{\Theta,\mathrm{s}}^{\mathrm{m}}(u)$ and $\underline{N}_{\Theta,\mathrm{r}}^{\mathrm{qr}}(u,t)$ in the SFD are defined as,

$$N_{\Theta,\mathrm{s}}^{\mathrm{m}}(u) = \frac{2n_{\mathrm{s}}\xi_{\mathrm{s,u}}}{\pi u p} \qquad (3.63) \qquad \underline{N}_{\Theta,\mathrm{r}}^{\mathrm{qr}}(u,t) = 2\mathrm{sinc}(\frac{u}{2})e^{-\mathrm{j}p\angle\theta_{\mathrm{r}}(t)} \qquad (3.64)$$

Within the rotor coordinate system xy, the linked magnetizing flux linkage vector for the stator $\vec{\psi}_{\mathrm{m}}^{\mathrm{s}}(t)$ and the rotor $\vec{\psi}_{\mathrm{m}}^{\mathrm{r,xy}}(t)$ can be expressed with the Clarke's Transformation shown in (3.65) and (3.66), with $m_{\mathrm{r}} = Q_{\mathrm{r}}/2p$.

$$\vec{\psi}_{\mathrm{m}}^{\mathrm{s}}(t) = \frac{2}{3} \begin{bmatrix} 1 & -\frac{1}{2} & -\frac{1}{2} \\ 0 & \frac{\sqrt{3}}{2} & -\frac{\sqrt{3}}{2} \end{bmatrix} \begin{bmatrix} \psi_{\mathrm{m}}^{1}(t) \\ \psi_{\mathrm{m}}^{2}(t) \\ \psi_{\mathrm{m}}^{3}(t) \end{bmatrix}$$

(3.65)

$$\vec{\psi}_{\mathrm{m}}^{\mathrm{r,xy}}(t) = \frac{2}{m_{\mathrm{r}}} \begin{bmatrix} \cos(0) & \sin(0) \\ \cos(\frac{\pi}{m_{\mathrm{r}}}) & \sin(\frac{\pi}{m_{\mathrm{r}}}) \\ \vdots & \vdots \\ \cos(\pi - \frac{\pi}{m_{\mathrm{r}}}) & \sin(\pi - \frac{\pi}{m_{\mathrm{r}}}) \end{bmatrix}^{\mathrm{T}} \begin{bmatrix} \psi_{\mathrm{m}}^{1,\mathrm{xy}}(t) \\ \psi_{\mathrm{m}}^{2,\mathrm{xy}}(t) \\ \vdots \\ \psi_{\mathrm{m}}^{\mathrm{qr,xy}}(t) \end{bmatrix}$$

(3.66)

3.1.2.8 Electromagnetic Force Calculation

The radial electromagnetic force excitation is the main effect for the radial force emissions and is defined in (3.67) [Sei92; MP09; GWC06; Bes08; BHK15].

$$F_{\mathrm{r}}(\alpha,t) = \frac{B_{\mathrm{air-gap}}^{2}(\alpha,t)}{2\mu_{0}} A_{\mathrm{surf}}, \quad F_{\mathrm{r}}(\alpha,t) \in \mathbb{R}$$

(3.67)

In order to complete the model in the SFD, the radial force is defined in (3.67). It consists of the convolution of the radial magnetic flux density $\underline{B}_{\mathrm{air-gap}}(u,t)$ with itself [Bis+17; Bis+16].

$$\underline{F}_{\mathrm{r}}(u,t) = \frac{\underline{B}_{\mathrm{air-gap}}(u,t) * \underline{B}_{\mathrm{air-gap}}(u,t)}{2\mu_{0}} A_{\mathrm{surf}}, \quad \underline{F}_{\mathrm{r}}(u,t) \in \mathbb{C}$$

(3.68)

This behavior is diagrammed in the temporal comparison of the run-ups of the radial forces in Fig. 3.37a and 3.37b, where additional harmonics occur due to this convolution.

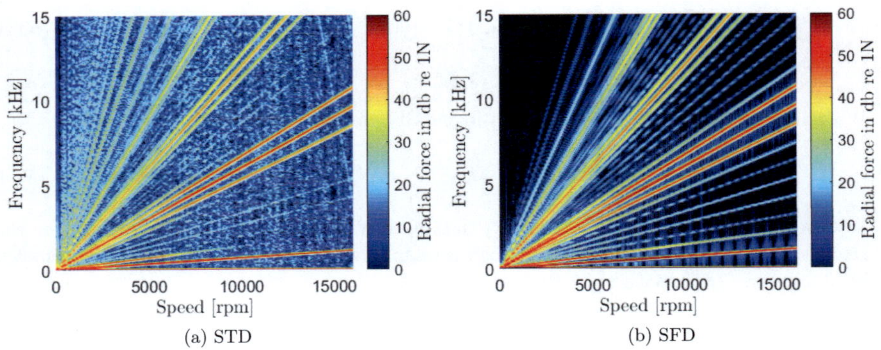

(a) STD (b) SFD

Figure 3.37: Run-up of the radial force in (a) STD and (b) SFD

3.2 Integrated Electromagnetic Simulation

The sequential electromagnetic simulation is a model in which the harmonics induction from the magnetic flux density into the stator coils and rotor bars is considered indirectly. With the findings of sequential simulation procedure, a model can be created including the electromagnetic flux density and its mutual effect on the stator and rotor currents. It is integrated in the system calculation as shown in Fig. 3.38. In comparison to the sequential electromagnetic simulation the force model calculates the non-sinusoidal flux linkages for the stator and rotor side as well, including all the spatial harmonics.

Earlier induction machine models only considered, e.g. the effects of saturation [Bis+01; ML92; MN83] or winding and slotting harmonics [Liw42; GMA99; Bes08] in space, while the combined effects of temporal harmonics from PWM voltage and magnetic saturation in an inverter fed induction machine is found in [Bes08; Mal00; BHK15], comparable to the sequential electromagnetic simulation form chapter 3, whereby all of these examples only use the STD. These approaches form the basis for this chapter and are refined for the integrated calculation procedure.

In order to create a model including all harmonic effects, it is prudent to focus on the magnetic circuit in the air-gap including the stator and rotor mmf, $\Theta_\mathrm{s}(\alpha,t)$ and $\Theta_\mathrm{r}(\alpha,t)$, the magnetic air-gap permeance $\Lambda_\mathrm{air-gap}(\alpha,t)$ and the magnetic flux density $B_\mathrm{air-gap}(\alpha,t)$, based on the calculation methods shown in the sequential electromagnetic simulation. Likewise important are the electrical quantities stator and rotor currents i_s and i_r, stator voltage u_s and magnetizing flux linkages, $\psi_\mathrm{m}^\mathrm{s}$ linked to stator winding and $\psi_\mathrm{m}^\mathrm{r}$ linked to rotor winding. A cause-effect relationship can be formed where firstly, the electrical currents energize the machine windings producing mmf, and secondly the magneto-motive force over the

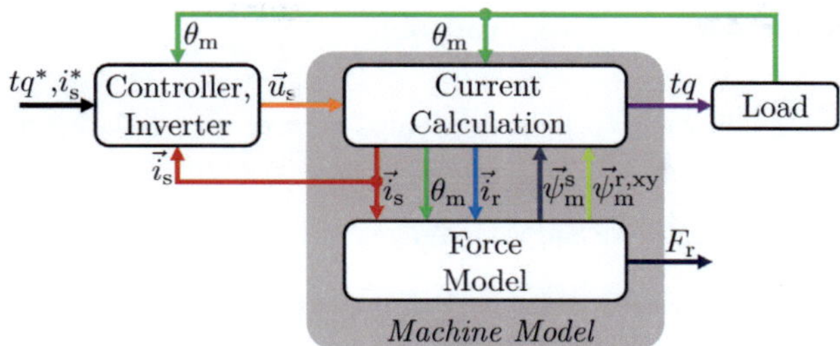

Figure 3.38: System integration of the integrated electromagnetic model

magnetic permeance creates the magnetic field including the harmonic components. This non-sinusoidal field links back again to the windings creating the magnetizing flux linkages. Lastly, the flux linkages carry the harmonic effects into their respective electrical circuits. A qualitative diagram of this causality chain is shown in Fig. 3.39.

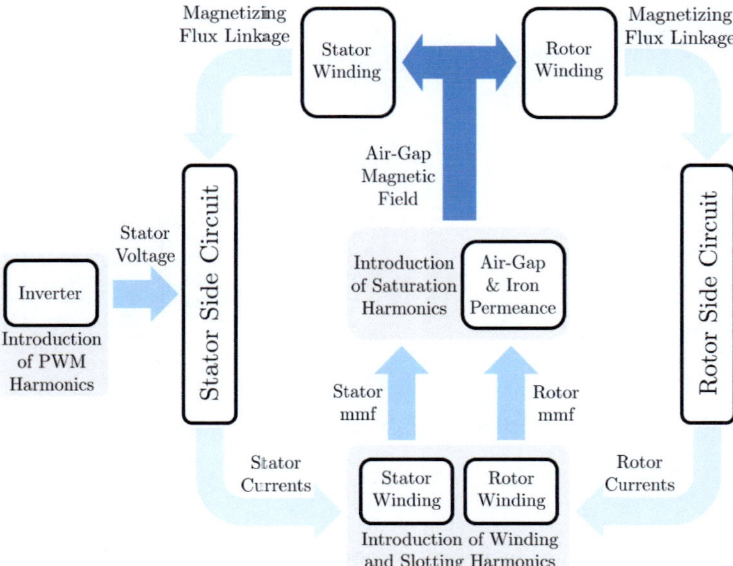

Figure 3.39: Qualitative diagram of the origins and propagation of machine harmonics

3.2.1 Modular Air-gap Flux Approach

The modular air-gap flux approach can be created in three stages relating to the harmonic effects:

- Stage 1: Leakage inclusive lumped air-gap model (LILAG)

- Stage 2: Leakage inclusive distributed air-gap model (LIDAG) without saturation

- Stage 3: Leakage inclusive distributed air-gap model (LIDAG) with saturation

First, as a lumped air-gap model is developed, the IRTF model approach is reshaped in order to receive the modular air-gap flux approach model, shown in Fig. 3.40. From this, a distributed air-gap model without saturation is derived. Lastly, the effect of saturation is included. A functionally comparable model for synchronous machines is derived in a similar process in [Boe+12]. However, from a technical point of view the models differ as the model in [Boe+12] relies on FEM or measurements [LBD14] for parametrization. The air-gap flux depends on the model structure related to the harmonic effects, that have to be included, shown in the following subsections.

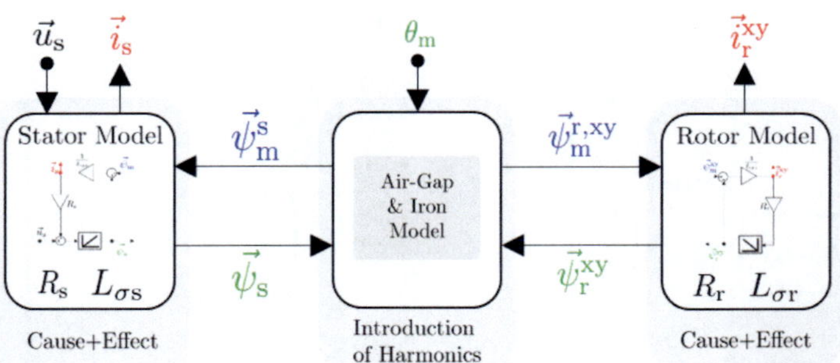

Figure 3.40: Schematic of the modular air-gap flux approach

3.2.1.1 Stage 1: Leakage Inclusive Lumped Air-gap Model (LILAG)

The first step towards modeling the various harmonic effects and their propagation over the stator and rotor system is to rearrange the IRTF fundamental flux model [DPV11], which forms the basis for this new model. Assuming a very small step size, the equations from the fundamental model, the IRTF model [DPV11], shown in (3.69a), can be written as (3.69b):

$$\vec{\psi}_{\mathrm{m}}(t_1) \approx L_{\mathrm{m}}(\vec{i}_{\mathrm{s}}(t_0) - \vec{i}_{\mathrm{r}}(t_0)), \qquad (3.69a) \qquad \vec{\psi}_{\mathrm{m}}(t_1) = \frac{\left(\frac{\vec{\psi}_{\mathrm{s}}(t_0)}{L_{\sigma,\mathrm{s}}} + \frac{\vec{\psi}_{\mathrm{r}}(t_0)}{L_{\sigma,\mathrm{r}}}\right) L_{\mathrm{m}}}{1 + \left(\frac{1}{L_{\sigma,\mathrm{s}}} + \frac{1}{L_{\sigma,\mathrm{r}}}\right) L_{\mathrm{m}}}. \qquad (3.69b)$$

Equation (3.69b) can be taken as LILAG with stator- and rotor-side flux linkages $\vec{\psi}_{\mathrm{s}}$ and $\vec{\psi}_{\mathrm{r}}$ as inputs and the magnetizing flux linkage $\vec{\psi}_{\mathrm{m}}$ as the output. Figure 3.41 shows the new LILAG, for which the input-output of the stator and rotor side are altered. In this model the respective currents are now the additional results of the stator and rotor side models.

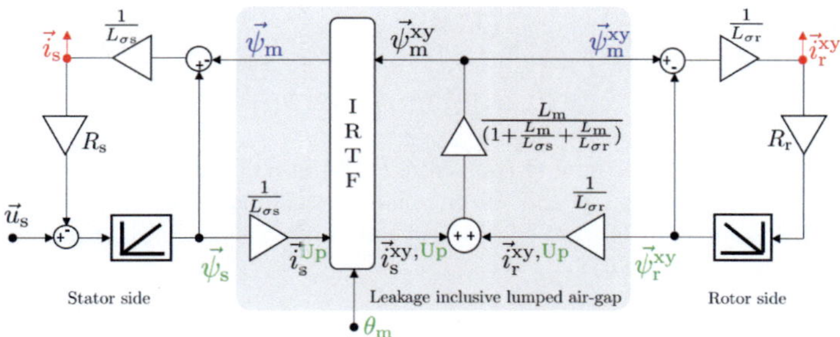

Figure 3.41: Stage 1: LILAG

The model, in Fig. 3.41, with a leakage inclusive lumped air-gap has no closed path which does not encounter an integral operator. This trivializes any numerical error propagation and avoids any algebraic loops. The magnetizing flux linkages $\vec{\psi}_{\mathrm{m}}$ and $\vec{\psi}_{\mathrm{m}}^{\mathrm{xy}}$ are still the inputs to the stator and rotor side, accounting for the induction from the other side of the air-gap respectively. Moreover, the continuous transfer functions of the stator and rotor side models with $\vec{\psi}_{\mathrm{s}}$ and $\vec{\psi}_{\mathrm{r}}$ as outputs are now devoid of any zeros and can be written as shown in (3.70a) and (3.70b).

$$\vec{\psi}_{\mathrm{s}} = \frac{L_{\sigma,\mathrm{s}}\vec{u}_{\mathrm{s}} + R_{\mathrm{s}}\vec{\psi}_{\mathrm{m}}}{R_{\mathrm{s}} + sL_{\sigma s}} \qquad (3.70a) \qquad \vec{\psi}_{\mathrm{r}}^{\mathrm{xy}} = \frac{R_{\mathrm{r}}\vec{\psi}_{\mathrm{m}}^{\mathrm{xy}}}{R_{\mathrm{r}} + sL_{\sigma r}} \qquad (3.70b)$$

As shown in this subsection, the fundamental model can be changed into a model structure, in order to adapt the air-gap function respectively. In this case, the air-gap function is equivalent to the fundamental behavior, neglecting the harmonics of the system.

3.2.1.2 Stage 2: Leakage Inclusive Distributed Air-gap Model (LIDAG) without Saturation

Keeping identical stator and rotor side models, the leakage inclusive lumped air-gap can now be modified to a LIDAG model. For didactic purposes, two new notations 'Up' and 'Down' are introduced. These refer to quantities originating directly from either the stator or rotor side flux linkages $\vec{\psi}_{\mathrm{s}}$ or $\vec{\psi}_{\mathrm{r}}$ as well as to quantities originating from the magnetizing flux linkages either $\vec{\psi}_{\mathrm{m}}^{\mathrm{s}}$ or $\vec{\psi}_{\mathrm{m}}^{\mathrm{r,xy}}$. Therefore, two virtual quantities can be defined for the stator currents $\vec{i}_{\mathrm{s}}^{\mathrm{Up}}$ and $\vec{i}_{\mathrm{s}}^{\mathrm{Down}}$ such as shown in (3.71a) - (3.71c) and similarly shown for the rotor in (3.72a) - (3.72c).

$$\vec{i}_{\mathrm{s}}^{\mathrm{Up}}(t) = \frac{\vec{\psi}_{\mathrm{s}}(t)}{L_{\sigma,\mathrm{s}}} \qquad (3.71\mathrm{a}) \qquad\qquad \vec{i}_{\mathrm{s}}^{\mathrm{Down}}(t) = \frac{\vec{\psi}_{\mathrm{m}}^{\mathrm{s}}(t)}{L_{\sigma,\mathrm{s}}} \qquad (3.71\mathrm{b})$$

$$\vec{i}_{\mathrm{s}}(t) = \vec{i}_{\mathrm{s}}^{\mathrm{Up}}(t) - \vec{i}_{\mathrm{s}}^{\mathrm{Down}}(t) \qquad (3.71\mathrm{c})$$

$$\vec{i}_{\mathrm{r}}^{\mathrm{xy,Up}}(t) = \frac{\vec{\psi}_{\mathrm{r}}^{\mathrm{xy}}(t)}{L_{\sigma,\mathrm{r}}} \qquad (3.72\mathrm{a}) \qquad\qquad \vec{i}_{\mathrm{r}}^{\mathrm{xy,Down}}(t) = \frac{\vec{\psi}_{\mathrm{m}}^{\mathrm{r,xy}}(t)}{L_{\sigma,\mathrm{r}}} \qquad (3.72\mathrm{b})$$

$$\vec{i}_{\mathrm{r}}^{\mathrm{xy}}(t) = \vec{i}_{\mathrm{r}}^{\mathrm{xy,Down}}(t) - \vec{i}_{\mathrm{r}}^{\mathrm{xy,Up}}(t) \qquad (3.72\mathrm{c})$$

Using (3.71c) and the stator mmf $\Theta_{\mathrm{s}}(\alpha,t)$ from (3.20) in STD or (3.21) in SFD, the stator mmf $\Theta_{\mathrm{s}}(\alpha,t)$ relating to the 'Up' and 'Down' quantities can be calculated, shown in (3.73a). Analogous the rotor mmf $\Theta_{\mathrm{r}}(\alpha,t)$ can be calculated, shown in (3.73b) using (3.72c) and (3.28) in STD or (3.29) in SFD.

$$\Theta_{\mathrm{s}}(\alpha,t) = \Theta_{\mathrm{s}}^{\mathrm{Up}}(\alpha,t) - \Theta_{\mathrm{s}}^{\mathrm{Down}}(\alpha,t) \qquad (3.73\mathrm{a})$$

$$\Theta_{\mathrm{r}}(\alpha,t) = \Theta_{\mathrm{r}}^{\mathrm{Down}}(\alpha,t) - \Theta_{\mathrm{r}}^{\mathrm{Up}}(\alpha,t) \qquad (3.73\mathrm{b})$$

Taking the air-gap permeance $\Lambda_{\mathrm{air-gap}}(\alpha,t)$ from (3.34) can be seen as a density $\mathrm{d}\Lambda/\mathrm{d}A$, the unsaturated radial magnetic density $B_{\mathrm{air-gap}}(\alpha,t)$ can be expressed as,

$$B_{\mathrm{air-gap}}(\alpha,t) = B_{\mathrm{air-gap}}^{\mathrm{Up}}(\alpha,t) - B_{\mathrm{air-gap}}^{\mathrm{Down}}(\alpha,t) \qquad (3.74\mathrm{a})$$

$$B_{\mathrm{air-gap}}^{\mathrm{Up}}(\alpha,t) = \left(\Theta_{\mathrm{s}}^{\mathrm{Up}}(\alpha,t) + \Theta_{\mathrm{r}}^{\mathrm{Up}}(\alpha,t)\right) \frac{\mathrm{d}\Lambda_{\mathrm{air-gap}}}{\mathrm{d}A_{\mathrm{surf}}} \qquad (3.74\mathrm{b})$$

$$B_{\mathrm{air-gap}}^{\mathrm{Down}}(\alpha,t) = \left(\Theta_{\mathrm{s}}^{\mathrm{Down}}(\alpha,t) + \Theta_{\mathrm{r}}^{\mathrm{Down}}(\alpha,t)\right) \frac{\mathrm{d}\Lambda_{\mathrm{air-gap}}}{\mathrm{d}A_{\mathrm{surf}}} \qquad (3.74\mathrm{c})$$

Using (3.74a), a cause-effect bifurcation can be made where 'Down' quantities are directly and solely caused by 'Up' quantities. For example, the stator single phase-turns linked magnetizing flux linkage from (3.75a) can be rewritten as shown in (3.75b), with r and l_{Fe} as the geometrical quantities for radius and length of the active material and $\Phi_{\mathrm{m}}^{\mathrm{m}}(t)$ as the flux. A similar bifurcation can be done for the rotor phase.

$$\psi_{\mathrm{m}}^{\mathrm{m}}(t) = \int N_{\Theta,\mathrm{s}}^{\mathrm{m}}(\alpha)\, \mathrm{d}\Phi_{\mathrm{m}}^{\mathrm{m}}(t)$$

$$= rl_{\mathrm{Fe}} \int_0^{2\pi} N_{\Theta,\mathrm{s}}^{\mathrm{m}}(\alpha) B_{\mathrm{air-gap}}(\alpha,t)\, \mathrm{d}\alpha \qquad (3.75a)$$

$$\psi_{\mathrm{m}}^{\mathrm{m}}(t) + rl_{\mathrm{Fe}} \int_0^{2\pi} N_{\Theta,\mathrm{s}}(\alpha) B_{\mathrm{air-gap}}^{\mathrm{Down}}(\alpha,t)\, \mathrm{d}\alpha$$

$$= rl_{\mathrm{Fe}} \int_0^{2\pi} N_{\Theta,\mathrm{s}}^{\mathrm{m}}(\alpha) B_{\mathrm{air-gap}}^{\mathrm{Up}}(\alpha,t)\, \mathrm{d}\alpha \qquad (3.75b)$$

A matrix Ψ is defined such that it contains the magnetizing flux linkages from all stator and rotor phases in their own reference frames and has dimensions $m+m_{\mathrm{r}}\times 1$,

$$\Psi = \begin{bmatrix} \psi_{\mathrm{m}}^1(t) & \psi_{\mathrm{m}}^2(t) & \psi_{\mathrm{m}}^3(t) & \psi_{\mathrm{m}}^{1,\mathrm{xy}}(t) & \cdots & \psi_{\mathrm{m}}^{\mathrm{qr,xy}}(t) \end{bmatrix}^{\mathrm{T}} \qquad (3.76)$$

Similarly, N is a matrix with the dimension $m+m_{\mathrm{r}}\times \mathrm{n}$ containing all stator and rotor phase turn functions in the stationary reference frame over the discrete 2π radial space, such that radial angle α now varies discretely with n as the spatial resolution,

$$N = \begin{bmatrix} N_{\Theta,\mathrm{s}}^1(\alpha) & N_{\Theta,\mathrm{s}}^2(\alpha) & N_{\Theta,\mathrm{s}}^3(\alpha) & N_{\Theta,\mathrm{r}}^1(\alpha,t) & \cdots & N_{\Theta,\mathrm{r}}^{\mathrm{qr}}(\alpha,t) \end{bmatrix}^{\mathrm{T}} \qquad (3.77)$$

Defining an inverse leakage inductance matrix C_σ as a phase modifier matrix of dimension $m+m_{\mathrm{r}}\times m+m_{\mathrm{r}}$ and using the previously defined permeance density, a matrix λ as a space modifier matrix can be defined over a discrete radial space with the dimensions $\mathrm{n}\times\mathrm{n}$,

$$C_\sigma = \begin{bmatrix} \frac{1}{L_{\sigma,\mathrm{s}}} & 0 & 0 & 0 & \cdots & 0 \\ 0 & \frac{1}{L_{\sigma,\mathrm{s}}} & 0 & 0 & \cdots & 0 \\ 0 & 0 & \frac{1}{L_{\sigma,\mathrm{s}}} & 0 & \cdots & 0 \\ 0 & 0 & 0 & \frac{1}{L_{\sigma,\mathrm{r}}} & \cdots & 0 \\ \vdots & \vdots & \vdots & \vdots & \ddots & \vdots \\ 0 & 0 & 0 & 0 & \cdots & \frac{1}{L_{\sigma,\mathrm{r}}} \end{bmatrix} \qquad (3.78)$$

$$\lambda = \begin{bmatrix} \frac{\mathrm{d}\Lambda(0,t)}{\mathrm{d}A} & 0 & \cdots & 0 \\ 0 & \frac{\mathrm{d}\Lambda(\frac{2\pi}{n},t)}{\mathrm{d}A} & \cdots & 0 \\ \vdots & \vdots & \ddots & \vdots \\ 0 & 0 & \cdots & \frac{\mathrm{d}\Lambda(2\pi,t)}{\mathrm{d}A} \end{bmatrix} \qquad (3.79)$$

Using (3.76) - (3.79), the $[B^{\mathrm{Down}}]^{\mathrm{T}}$ matrix in (3.75b) can be expanded as shown in (3.80a). Taking $2\pi r l_{\mathrm{Fe}}/\mathrm{n}$ in (3.75b) as the area of a thin strip A_Δ on the curved surface and generalizing for all stator and rotor phases, (3.80b) can be defined. Here I is an identity matrix of dimension $[m+m_{\mathrm{r}}, m+m_{\mathrm{r}}]$ premultiplied to Ψ and can be rearranged to (3.80c). From (3.80c), the LIDAG is created. Taking the definition of $B_{\mathrm{air-gap}}^{\mathrm{Up}}$ from (3.74b) and keeping the stator and rotor side models identical to (3.70a) and (3.70b), the model can be seen in

Fig. 3.42. The respective magnetizing flux linkages space vectors $\vec{\psi}_{\mathrm{m}}^{\mathrm{s}}$ and $\vec{\psi}_{\mathrm{m}}^{\mathrm{r,xy}}$ can be taken directly from Ψ.

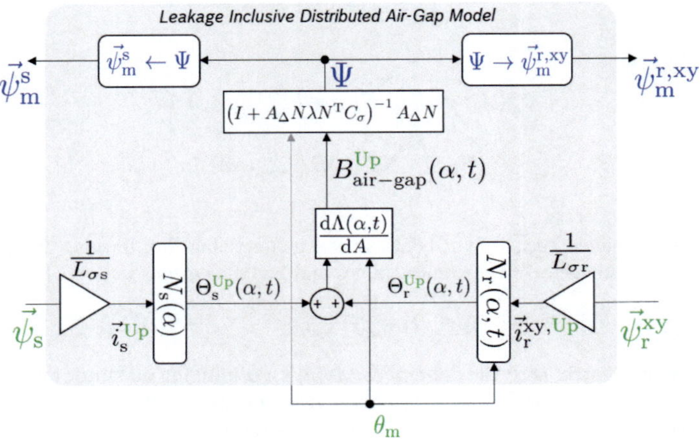

Figure 3.42: Stage 2: LIDAG without saturation

$$\left[B_{\text{air-gap}}^{\text{Down}}\right]^{\mathrm{T}} = \lambda N^{\mathrm{T}} C_\sigma \Psi \tag{3.80a}$$

$$I\Psi + A_\Delta N \lambda N^{\mathrm{T}} C_\sigma \Psi = A_\Delta N \left[B_{\text{air-gap}}^{\mathrm{Up}}\right]^{\mathrm{T}} \tag{3.80b}$$

$$\Psi = \left[I + A_\Delta N \lambda N^{\mathrm{T}} C_\sigma\right]^{-1} A_\Delta N \left[B_{\text{air-gap}}^{\mathrm{Up}}\right]^{\mathrm{T}} \tag{3.80c}$$

If a matrix Ψ is defined similar to Ψ but for all stator and rotor phase flux linkages ψ_{s} and $\psi_{\mathrm{r}}^{\mathrm{xy}}$, B^{Up} can also be similarly expanded as in (3.81a). Thus relating Ψ and Ψ can be written as shown in (3.81b). Now, if two machine matrices G and M are defined such that they constitute the information about the geometric configuration of the machine, G and M can be defined as (3.81c) and (3.81d).

$$\left[B_{\text{air-gap}}^{\mathrm{Up}}\right]^{\mathrm{T}} = \lambda N^{\mathrm{T}} C_\sigma \Psi \tag{3.81a}$$

$$\Psi = \left[I + A_\Delta N \lambda N^{\mathrm{T}} C_\sigma\right]^{-1} A_\Delta N \lambda N^{\mathrm{T}} C_\sigma \Psi \tag{3.81b}$$

$$G = A_\Delta N \tag{3.81c} \qquad\qquad M = \lambda N^{\mathrm{T}} C_\sigma \tag{3.81d}$$

The matrix G can be seen as the equivalent linkage matrix and M can be seen as the magnetization matrix. With (3.81c) and (3.81d) the input-output relationship of the LIDAG model without saturation can be reduced to (3.82a) and can be rearranged using (3.74a), (3.82a) as shown in (3.82b).

$$\Psi = [I + GM]^{-1} GM\Psi \tag{3.82a}$$
$$\Psi = GM\Psi - GM\Psi = G[B_{\text{air-gap}}(\alpha,t)] \tag{3.82b}$$

As shown in this subsection, the LILAG of section 3.2.1.1 can be changed including the spatial magnetic flux density in the air-gap of the induction machine. In this case, the air-gap function includes the spatial winding and slotting harmonics, which cause non-sinusoidal flux linkages and subsequently stator and rotor currents including the winding and slotting machine harmonics.

3.2.1.3 Stage 3: Leakage Inclusive Distributed Air-gap Model (LIDAG) with Saturation

When modeling the effects of saturation, the non-conservative nature of the magnetic field lines become more prominent. This is accounted in the model by separating the stator and rotor systems such that the self and mutual interactions on either side of the air gap experience different saturation functions $f_{\text{sat,s}}$ and $f_{\text{sat,r}}$ akin to stator yoke and rotor core saturation, shown in Fig. 3.43. Both curves are unitized, non-linear, absolute decreasing functions over mmf magnitude, generated using a general B-H curve of a magnetic material. The same air-gap mmf is applied to estimate the saturation scaling factors $k_{\text{sat,s}}$ and $k_{\text{sat,r}}$ varying over time and radial space

$$k_{\text{sat,s}}(\alpha,t) = f_{\text{sat,s}}\left(\Theta_{\text{s}}(\alpha,t) + \Theta_{\text{r}}(\alpha,t)\right) \tag{3.83a}$$
$$k_{\text{sat,r}}(\alpha,t) = f_{\text{sat,r}}\left(\Theta_{\text{s}}(\alpha,t) + \Theta_{\text{r}}(\alpha,t)\right) \tag{3.83b}$$

Using (3.83a), (3.83b) and (3.79), the saturated permeance matrices of the stator and rotor systems can be written as,

$$\lambda_{\text{sat,s}} = k_{\text{sat,s}}(\alpha,t)\lambda \tag{3.84a} \qquad \lambda_{\text{sat,r}} = k_{\text{sat,r}}(\alpha,t)\lambda \tag{3.84b}$$

Bifurcating the turns functions N_{s} and N_{r} and inverse leakage inductance matrices $C_{\sigma,\text{s}}$ and $C_{\sigma,\text{r}}$ as

$$N_{\text{s}} = \left[N^1_{\Theta,\text{s}}(\alpha) \; N^2_{\Theta,\text{s}}(\alpha) \; N^3_{\Theta,\text{s}}(\alpha)\right]^{\text{T}} \tag{3.85a}$$
$$N_{\text{r}} = \left[N^1_{\Theta,\text{r}}(\alpha,t) \; \cdots \; N^{\text{qr}}_{\Theta,\text{r}}(\alpha,t)\right]^{\text{T}} \tag{3.85b}$$

$$C_{\sigma,\mathrm{s}} = \begin{bmatrix} \frac{1}{L_{\sigma s}} & 0 & 0 \\ 0 & \frac{1}{L_{\sigma,\mathrm{s}}} & 0 \\ 0 & 0 & \frac{1}{L_{\sigma,\mathrm{s}}} \end{bmatrix} \quad (3.86a) \qquad C_{\sigma,\mathrm{r}} = \begin{bmatrix} \frac{1}{L_{\sigma,\mathrm{r}}} & \cdots & 0 \\ \vdots & \ddots & \vdots \\ 0 & \cdots & \frac{1}{L_{\sigma,\mathrm{r}}} \end{bmatrix} \quad (3.86b)$$

The new linkage and magnetizing matrices for the stator and rotor systems can be written as,

$$M_{\mathrm{sat,ss}} = \lambda_{\mathrm{sat,s}} N_{\mathrm{s}}^{\mathrm{T}} C_{\sigma,\mathrm{s}} \tag{3.87c}$$

$$G_{\mathrm{s}} = A_{\Delta} N_{\mathrm{s}} \tag{3.87a} \qquad M_{\mathrm{sat,sr}} = \lambda_{\mathrm{sat,s}} N_{\mathrm{r}}^{\mathrm{T}} C_{\sigma,\mathrm{r}} \tag{3.87d}$$

$$G_{\mathrm{r}} = A_{\Delta} N_{\mathrm{r}} \tag{3.87b} \qquad M_{\mathrm{sat,rs}} = \lambda_{\mathrm{sat,r}} N_{\mathrm{s}}^{\mathrm{T}} C_{\sigma,\mathrm{s}} \tag{3.87e}$$

$$M_{\mathrm{sat,rr}} = \lambda_{\mathrm{sat,r}} N_{\mathrm{r}}^{\mathrm{T}} C_{\sigma,\mathrm{r}} \tag{3.87f}$$

Referring to (3.82b) the complete system with saturation can be written as,

$$\Psi = GM_{\mathrm{sat}}\Psi - GM_{\mathrm{sat}}\Psi \tag{3.88}$$

where,

$$GM_{\mathrm{sat}} = \begin{bmatrix} G_{\mathrm{s}} M_{\mathrm{sat,ss}} & G_{\mathrm{s}} M_{\mathrm{sat,sr}} \\ G_{\mathrm{r}} M_{\mathrm{sat,rs}} & G_{\mathrm{r}} M_{\mathrm{sat,rr}} \end{bmatrix} \tag{3.89}$$

The matrices $G_{\mathrm{s}} M_{\mathrm{sat,ss}}$ and $G_{\mathrm{s}} M_{\mathrm{sat,sr}}$ are the self and mutual interaction over the stator saturation function $f_{\mathrm{sat,s}}$ and similarly $G_{\mathrm{s}} M_{\mathrm{sat,rr}}$ and $G_{\mathrm{s}} M_{\mathrm{sat,rs}}$ over the rotor saturation function $f_{\mathrm{sat,r}}$. The final magnetization flux linkage matrix Ψ can be written as,

$$\Psi = [I + GM_{\mathrm{sat}}]^{-1} GM_{\mathrm{sat}}\Psi \tag{3.90}$$

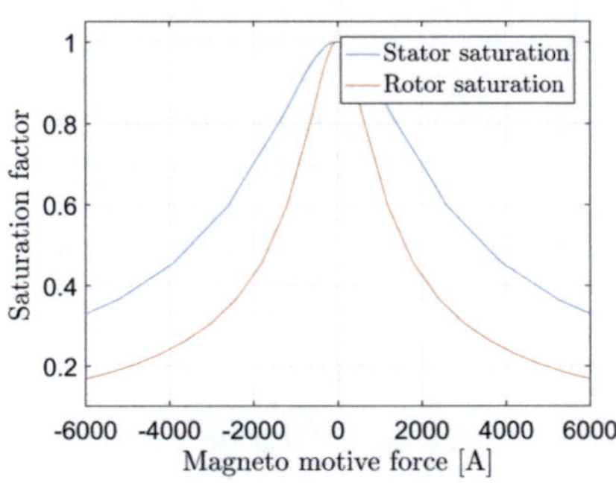

Figure 3.43: Saturation factor for stator yoke and rotor core saturation

The LIDAG model is represented by (3.90) and together with (3.70a) and (3.70b) it represents the machine model. The resultant model includes the effects of slotting, teeth placement and saturation. The input-output relationship of Ψ and Ψ for the stator and rotor side is inherently stable in this model. The current space-vectors come out of this model as a by-product and can be calculated from (3.71c) and (3.72c). As the model follows the causality chain with respect to harmonics, the effects of PWM harmonics introduced by the stator voltage can also be modeled.

3.2.2 Identification of Occurring Orders

As already shown in section 3.1, and completed in section 3.2, the spatial and temporal components of the magnetic flux density cause the NVH-behavior of the induction machine. Summarizing the occurring harmonics of the magnetic flux density and its components, the spatial and temporal effects can be explained as follows [Wil71; VB94; GWC06; MP09; Bis+16; Bis+17; Sei92].

Spatial electromagnetic orders

In the stator mmf $\Theta_\mathrm{s}(\alpha,t)$, energized by the number of fundamental phase currents m, specific spatial harmonic orders ν are observed. These spatial orders can be categorized into winding harmonics ν_winding, which occur due to a finite order of stator phases m distributed of 2π radial space and slotting harmonics ν_slot. These in turn occur due to $\frac{Q_\mathrm{s}}{p}$ number of discrete stator slots Q_s per pole pair p in the radial space, as shown in (3.91a) and (3.91b).

$$\nu_\mathrm{winding} = 1 \pm 2m_\mathrm{s}k_\nu \text{ , where } k_\nu \in \mathbb{Z} \tag{3.91a}$$

$$\nu_\mathrm{slot} = 1 \pm \frac{Q_\mathrm{s}k_\nu}{p} \text{ , where } k_\nu \in \mathbb{Z} \tag{3.91b}$$

The spatial rotor winding harmonic orders μ_winding are directly affected by the spatial stator winding orders ν_winding due to the magnetic induction. The orders can be found by taking one spatial harmonic period $\frac{2\pi}{\nu}$ of a spatial stator order and calculating the division of this radial space by the discrete rotor bars Q_r over the pole pair number p. The occurring winding harmonics can be written as (3.92a) and the slotting harmonics as (3.92b).

$$\mu_\mathrm{winding} = \left(\nu_\mathrm{winding} \pm \frac{Q_\mathrm{r}k_\mu}{p} \right) \text{ , where } k_\mu \in \mathbb{Z} \tag{3.92a}$$

$$\mu_\mathrm{s.ot} = \left(1 \pm \frac{Q_\mathrm{r}k_\mu}{p} \right) \text{ , where } k_\mu \in \mathbb{Z} \tag{3.92b}$$

Spatial orders, which occur due to the mutual interaction of stator and rotor side, are defined in (3.93).

$$\nu_{\text{int}} = \nu \pm \mu \pm 1 \tag{3.93}$$

The saturation of the magnetic permeance of the stator and rotor side affecting the magnetic flux density in space can be taken as $\nu_{\text{sat}} = \mu_{\text{sat}}$ and are shown in (3.94a) for the interaction of spatial stator orders. Equation (3.94b) displays the interaction of spatial rotor orders.

$$\nu' = \nu + \nu_{\text{sat}} \text{ , where } \nu_{\text{sat}} = 2k_{\text{sat},\nu} \text{ , } k_{\text{sat},\nu} \in \mathbb{Z} \tag{3.94a}$$
$$\mu' = \mu + \mu_{\text{sat}} \text{ , where } \mu_{\text{sat}} = 2k_{\text{sat},\mu} \text{ , } k_{\text{sat},\mu} \in \mathbb{Z} \tag{3.94b}$$

Temporal electromagnetic orders

With this spatial information, the occurring temporal stator and rotor current orders, are defined in (3.95a). Equation (3.95b) represents the machine slip s. As the effective saturation wave moves in the stator and rotor systems with fundamental stator and slip frequency, the respective saturation orders can be directly incorporated on top of the system. The temporal harmonic orders of the stator flux linkage are the same as that of the stator current. The torque is the convolution of the current and flux linkage. Therefore, its temporal orders can be written as shown in (3.95c).

$$\tau_{\text{s}}\Big|_{\tau_{\text{s}} \neq 3,6,9...} = s + \mu(1-s) \pm \nu_{\text{sat}} \tag{3.95a} \qquad \tau_{\text{r}} = \frac{1 - \nu(1-s)}{s} \pm \mu_{\text{sat}} \tag{3.95b}$$

$$\tau_{\text{tq}} = \pm \tau_{\text{s}} \pm \tau_{\text{r}} \tag{3.95c}$$

Spatial mechanical orders

In addition to the electromagnetic effects, the mechanical slotting orders itself have a direct influence on the magnetic flux density, subsequently the radial forces, as well as the stator and rotor currents. The spatial mechanical stator and rotor slotting harmonics, as well as their interaction, can be described as shown in (3.96a),(3.96b) and (3.97).

$$\nu^* = \pm \frac{Q_{\text{s}}}{p} k_{\nu^*}, \text{ with } k_{\nu^*} \in \mathbb{Z} \tag{3.96a} \qquad \mu^* = \pm \frac{Q_{\text{r}}}{p} k_{\mu^*}, \text{ with } k_{\mu^*} \in \mathbb{Z} \tag{3.96b}$$

$$\nu^*_{\text{int}} = \nu^* + \mu^* \tag{3.97}$$

In case of eccentricity, there are additional spatial harmonics occurring as,

$$\nu^* = \nu^* \pm \frac{1}{p} \tag{3.98a} \qquad \mu^* = \mu^* \pm \frac{1}{p} \tag{3.98b}$$

Temporal mechanical orders

The temporal harmonics occurring due to the induction from the mechanical harmonics in the flux linkages into the stator and rotor currents are shown in (3.99a) and (3.99b). The torque harmonics are calculated similar to the electromagnetic case, shown in (3.99c). The orders of the electromagnetic and mechanical effects show a small difference and are often taken together as slotting harmonics.

$$\tau_s^*\Big|_{\tau_s^* \neq 3,6,9\ldots} = \mu^*(1-s) \pm \nu_{\text{sat}} \qquad (3.99a) \qquad\qquad \tau_r^* = \frac{-\mu^*(1-s)}{s} \pm \mu_{\text{sat}} \qquad (3.99b)$$

$$\tau_{\text{tq}}^* = \pm\tau_s^* \pm \tau_r^* \qquad\qquad (3.99c)$$

Dependency on machine poles

Including the number of poles, the spatial and temporal orders are multiples of the machine pole pair number p, as

$$\tilde{\nu} = \nu p \qquad (3.100a) \qquad\qquad \tilde{\mu} = \mu p \qquad (3.100b) \qquad\qquad \tilde{\tau} = \tau p \qquad (3.100c)$$

For the noise calculation it is important to mention, that the calculation of the magnetic flux density consists of a convolution with itself. This causes multitude orders with different amplitudes, complicating the finding of its origins. Nevertheless, the prominent harmonics content of induction machines with open slots can be defined as the fundamental, slotting and saturation harmonics, as well as the interaction of them.

Interaction with supply voltage harmonics

Feeding the induction machine with non-sinusoidal voltages produce additional temporal harmonics in the machine. As shown in section 2.3.2, the inverter control schemes cause different harmonic content. The final temporal orders h_s and h_r induced from the air-gap flux linkages into the stator and rotor currents, including the effects of winding, slotting and saturation and its interaction with PWM harmonics can now be condensed into the following expressions, with $h_{\text{PWM}} = f_{\text{PWM}}/f_{\text{elec}}$,

$$h_s = h_r s + \mu(1-s) \pm \nu_{\text{sat}} \cup h_{\text{PWM}} \qquad\qquad (3.101a)$$

$$h_r = \frac{h_s}{3} - \frac{\nu(1-s)}{s} \pm \mu_{\text{sat}}. \qquad\qquad (3.101b)$$

3.3 Validation and Comparison

This section draws a comparison between the different implementation methods and electromagnetic FEM results fed with sinusoidal voltages. Further more, this section compares the validation with test bench measurement results fed with PWM-Voltage supply. First, the stator and rotor currents, electromagnetic torque, as well as the magnetic flux density and subsequently the radial force from the fundamental model, sequential electromagnetic and integrated electromagnetic model are compared. In the next step, the integrated electromagnetic model is compared to FEM. Finally, the integrated electromagnetic model is validated with currents and torque measured at a test bench.

3.3.1 Validation with Simulations

In order to compare the sequential and integrated electromagnetic model to the FEM model, operating points (OPs) are defined, shown in table 3.1. The fundamental values describe the machine operation result for the fundamental model.

Table 3.1: Operating points (OPs) for comparison

Operating points (OPs)	1	2
Speed [rpm]	1000	6000
Phase voltage [V] rms	18.63	65.27
Stator frequency [Hz]	39.547	203.139
Fundamental torque [Nm]	133	48
Fundamental d-axis current I_d [A] rms	187.18	100.55
Fundamental q-axis current I_q [A] rms	407.41	173.1

3.3.1.1 Comparison between Sequential and Integrated Electromagnetic Models

Firstly, this section compares the sequential electromagnetic model coupled with the fundamental model and the extended fundamental model, including the extension of the rotor bar current calculation, implementing additional harmonics into the rotor mmf. Secondly the sequential model is compared to the integrated electromagnetic LIDAG without saturation waves. For the comparison, the saturation is included into the fundamental parameters of the model, changing the inductances.

Stator current

The stator current U-phases are shown for OP1 in Fig. 3.44a in TTD and Fig. 3.44b in TFD as well as for OP2 in Fig. 3.45a in TTD and Fig. 3.45b in TFD. Since the fundamental model and the extended fundamental model have both the fundamental stator current without harmonics, the position and amplitude of the 1st Order is compared to the LIDAG. The position of the stator currents in OP1 is the same, where in OP2 the position of the LIDAG has a small deviation due to the slotting effect and subsequently the induced voltages. The fundamental amplitudes of both OPs show a small difference in the LIDAG as well, caused by these effects.

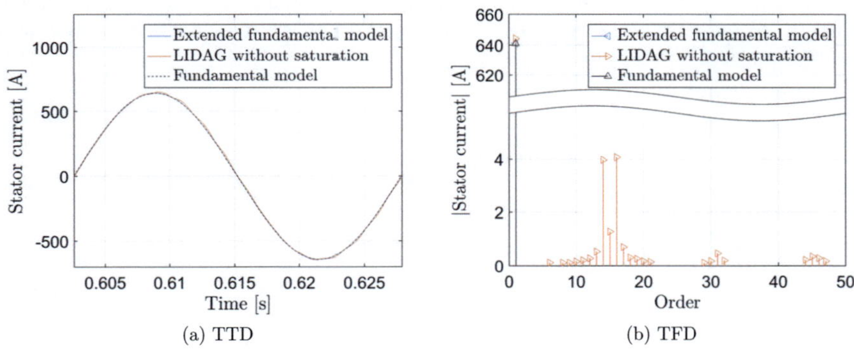

(a) TTD (b) TFD

Figure 3.44: Stator phase currents for OP1 in (a) TTD and (b) TFD

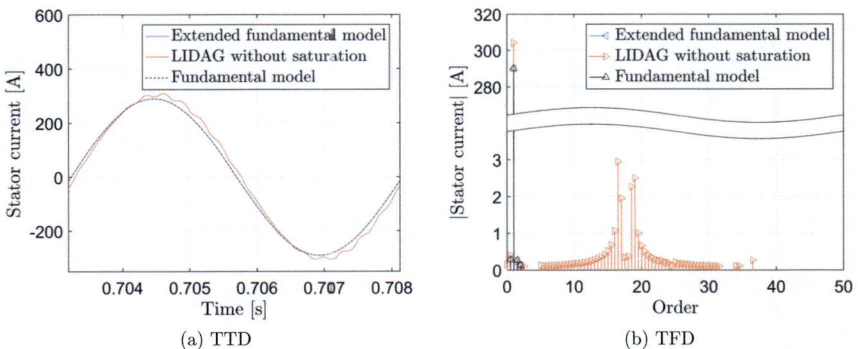

(a) TTD (b) TFD

Figure 3.45: Stator phase currents for OP2 in (a) TTD and (b) TFD

Rotor current

The rotor bar currents are shown in TTD and TFD in Fig. 3.46a and Fig. 3.46b for operating point 1 and in Fig. 3.47a and Fig. 3.47b for operating point 2. The currents from the fundamental model are still sinusoidal. The rotor bar currents from the extended fundamental model have additional harmonics, relating to the harmonic resistances and inductances of the extended rotor circuits. The rotor bar currents form the LIDAG are depending on the induced voltages from the flux linkage and with it from the magnetic flux distribution directly. It is shown, that the orders are matching exactly. The amplitudes of the LIDAG are not completely matching, because the slotting of the flux distributed model is assessed to be higher compared to the extended fundamental model. With this effect, the induction and subsequently the fundamental and the harmonics, as well as the phase angle, are not exactly the same. By neglecting the spatial saturation waves, the induction is reduced to the fundamental and the slotting effects.

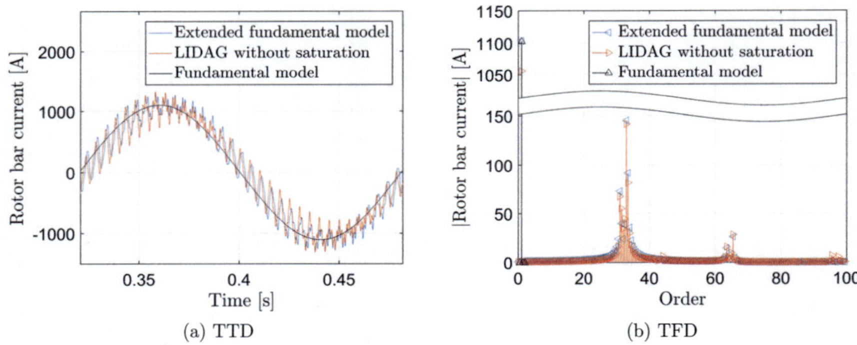

(a) TTD

(b) TFD

Figure 3.46: Rotor bar currents for OP1 in (a) TTD and (b) TFD

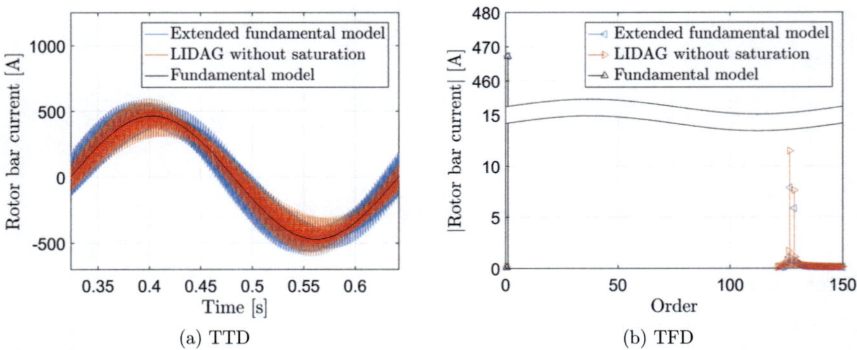

(a) TTD

(b) TFD

Figure 3.47: Rotor bar currents for OP2 in (a) TTD and (b) TFD

Electromagnetic torque

The electromagnetic torque is shown for OP1 and OP2 in TTD and TFD in Fig. 3.48a and Figr. 3.48b, as well as in Fig. 3.49a and Fig. 3.49b. The torque excitation of the fundamental and extended fundamental fed models shows the fundamental behavior, whereas the LIDAG directly calculates the torque harmonics of the spatial distributed flux density. Because of the already explained slotting effect, in section 3.1, the fundamental torque is rated to be lower for the LIDAG, compared to the fundamental model. Harmonics, which occur here are due to the slotting effect, without the neglected spatial saturation waves.

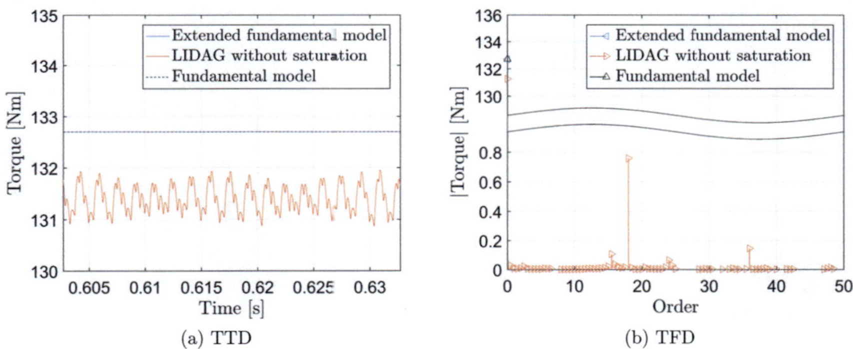

(a) TTD

(b) TFD

Figure 3.48: Electromagnetic torque for OP1 in (a) TTD and (b) TFD

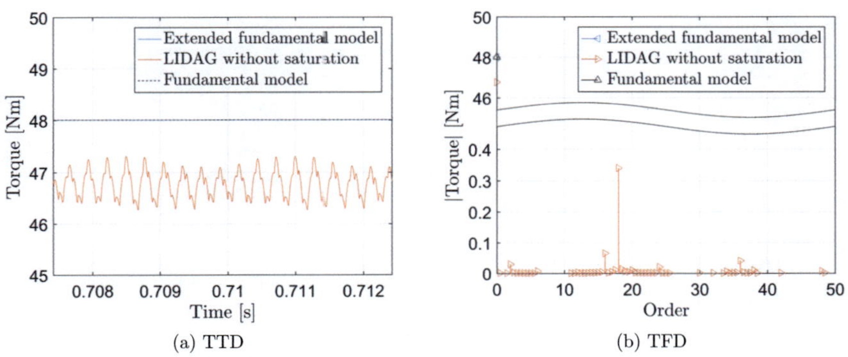

(a) TTD

(b) TFD

Figure 3.49: Electromagnetic torque for OP2 in (a) TTD and (b) TFD

Magneto motive force

The magneto motive forces in the temporal time domain beneath on stator tooth are shown in Fig. 3.50a and Fig. 3.50b. The stator and rotor mmfs, Θ_s and Θ_r, have the same phase shift like their excitation current. When comparing the rotor mmf Θ_r, the position of the rotor currents show marginal differences, due to the small phase shift of the fundamental and harmonics in the rotor current excitation. The rotor mmf excited by the sequential electromagnetic model and fed with the fundamental rotor current shows the most rectangular behavior. Adding harmonics to the rotor current smoothes the rotor mmf and changes the harmonics content of the summed mmf Θ ($\Theta = \Theta_s + \Theta_r$). The addition of stator mmf and rotor mmf does not change the occurring harmonics. In fact, the occurring harmonics are added up.

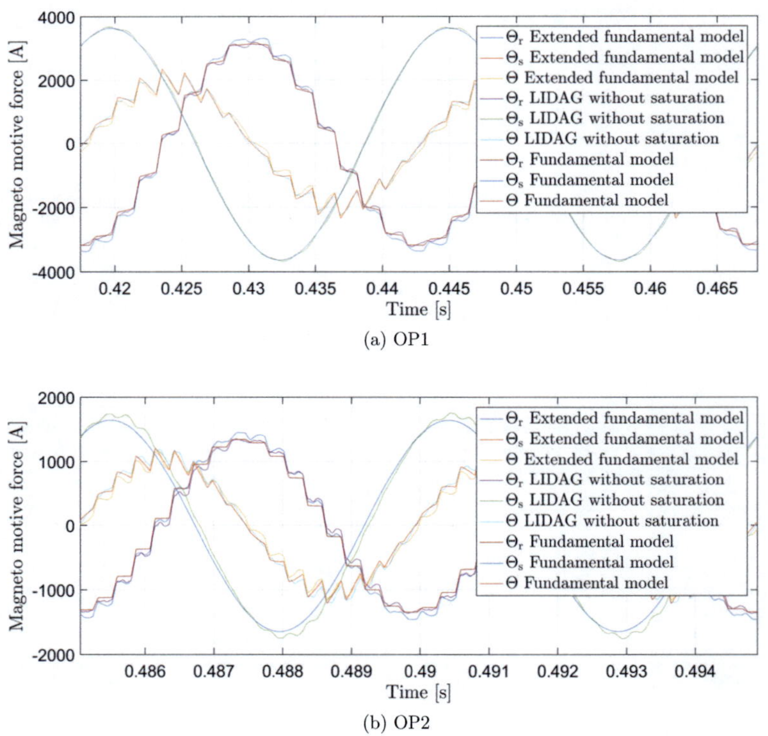

(a) OP1

(b) OP2

Figure 3.50: Temporal comparison of mmfs beneath one stator tooth for (a) OP1 and (b) OP2 in TTD

Magnetic flux density

The comparison of the spatial magnetic flux densities in the spatial time domain are shown in Fig. 3.51a and Fig. 3.51b for the operating point 1 and operating point 2. The position of the compared flux densities of OP1 matches, whereby the position of the LIDAG of OP2 shows a marginal phase shift. Differences occur due to the marginal differences in the stator and rotor current harmonics as well as the induced voltage, which affects the height of the tooth and the fundamental wave.

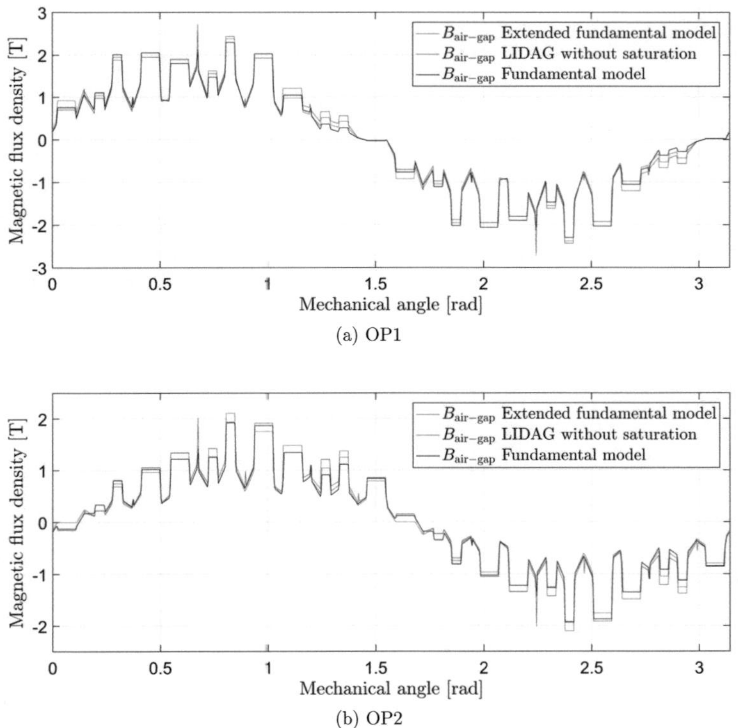

(a) OP1

(b) OP2

Figure 3.51: Spatial comparison of magnetic flux densities at one time step for (a) OP1 and (b) OP2 in STD

In the spatial frequency domain the differences are represented more clearly. The spatial harmonics content shows spatially increased differences in the lower spatial order. Fig. 3.52a

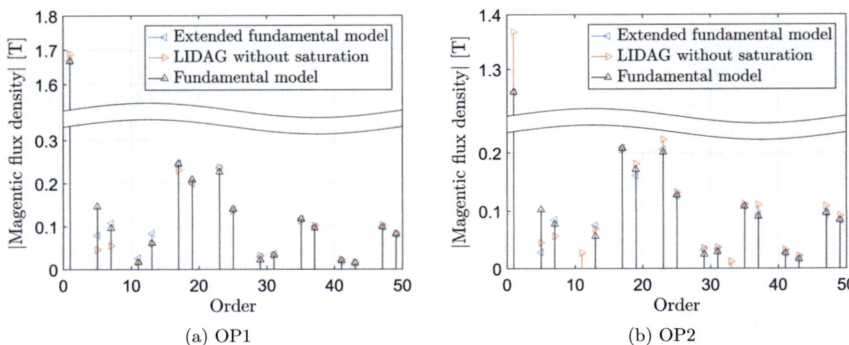

(a) OP1

(b) OP2

Figure 3.52: Spatial comparison of magnetic flux densities at one time step for (a) OP1 and (b) OP2 in SFD

shows the SFD at one time step for OP1, where the 5th and 7th order shows the biggest changes in the extended fundamental model and LIDAG in comparison to the fundamental model, neglecting the temporal harmonics effects. This is also the case in Fig. 3.52b for OP2, where the 1st order of the LIDAG reveals a major difference. This distinction is explained by the phase shift of the stator current and subsequently the phase shift of the stator mmf to the rotor mmf, increasing the fundamental wave.

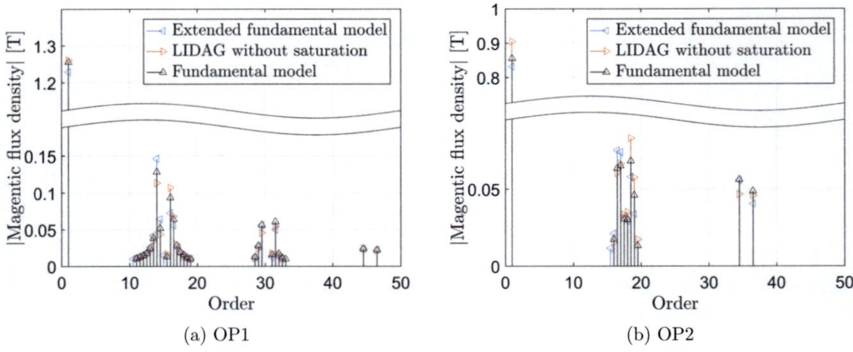

(a) OP1

(b) OP2

Figure 3.53: Temporal comparison of magnetic flux densities beneath one stator tooth for (a) OP1 and (b) OP2 in TFD

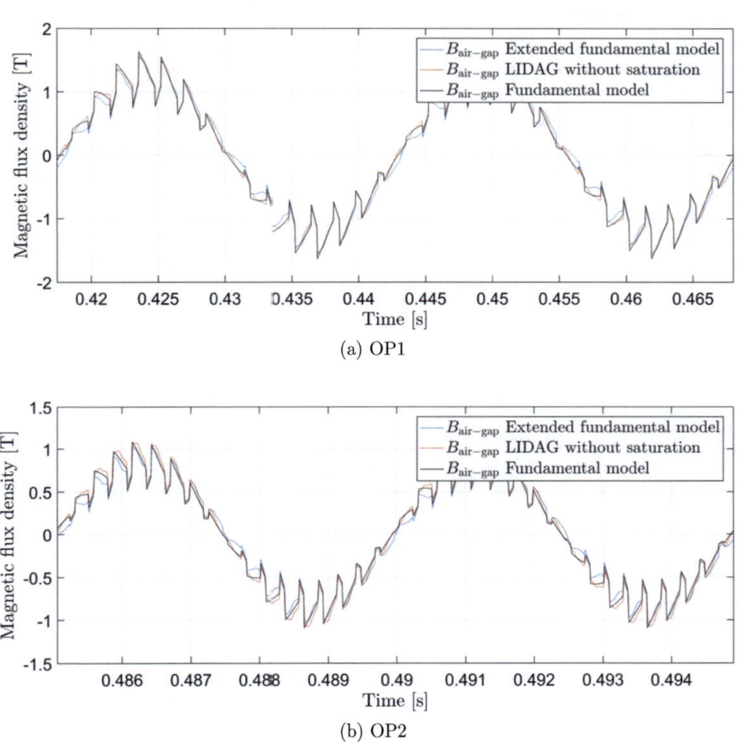

(a) OP1

(b) OP2

Figure 3.54: Temporal comparison of magnetic flux densities beneath one stator tooth for (a) OP1 and (b) OP2 in TTD

The temporal magnetic flux densities are compared beneath one stator tooth in the temporal time domain in Fig. 3.54a and Fig. 3.54b for the two operating points. In both cases, the rotor tooth surface structure becomes less rectangular, caused by the smoothed rotor mmfs due to the non-sinusoidal rotor currents. The phase shift of the stator current of the LIDAG changes the magnetic flux density in OP2 less than in the spatial comparison.

This can also be observed for the temporal frequency domain in Fig. 3.53a and Fig. 3.53b for OP1 and OP2. The differences of the magnitudes are very small in comparison to the spatial analysis. The complete harmonics content is led by the slotting effects, that are matching.

Radial force

The occurring radial forces, due to the radial magnetic flux densities, are shown for OP1 and OP2 in Fig. 3.55a and in Fig. 3.55b for one time step in the spatial time domain. In comparison to the magnetic flux densities, the differences between the models are increased and decreased, depending on the harmonics distribution. This is due to the squaring of the magnetic flux densities, which is a mathematical convolution in the frequency domain.

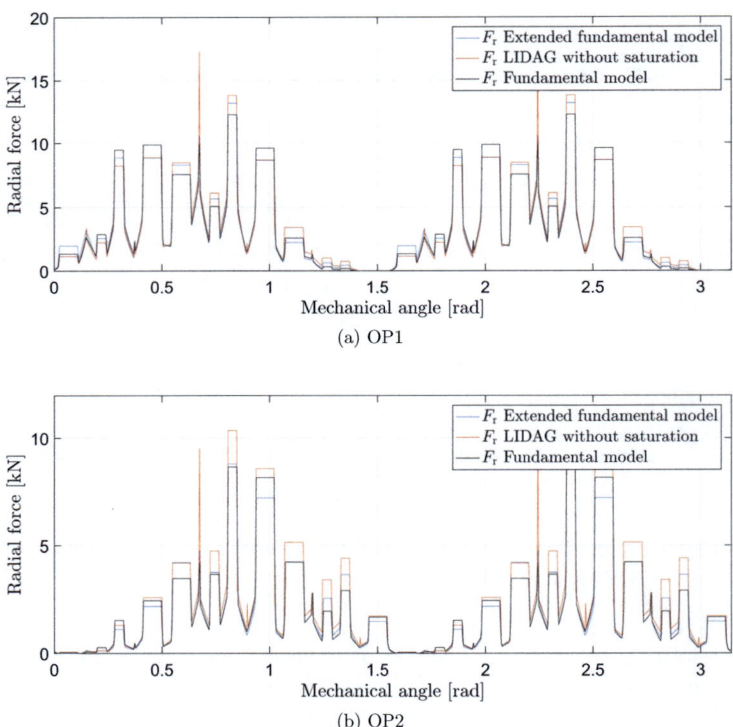

Figure 3.55: Spatial comparison of radial forces at one time step for (a) OP1 and (b) OP2 in STD

As it is already shown in the magnetic flux densities, the harmonics of lower orders like the 5th and 7th are differing from the fundamental model excited currents, shown in Fig. 3.56a and Fig. 3.56b in the SFD. Especially operating point 2 shows increased differences

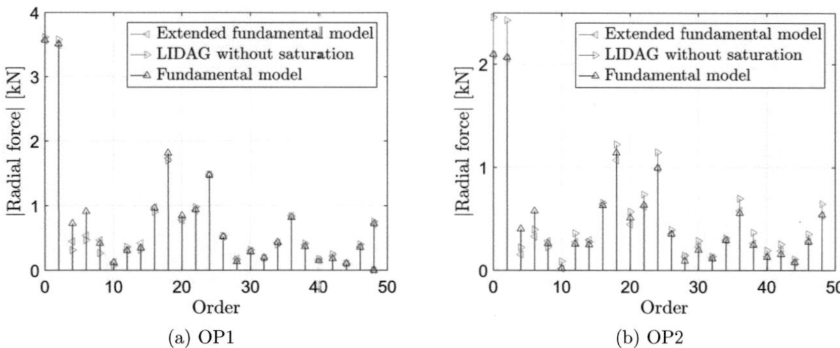

Figure 3.56: Spatial comparison of radial forces at one time step for (a) OP1 and (b) OP2 in SFD

in the 0th and 2nd spatial order. These effects can be attributed to the convolution of the increased 1st order of the magnetic flux density with itself.

The temporal comparison of the radial forces beneath one stator tooth is shown for these OPs in Fig. 3.57a and Fig. 3.57b for the TFD and in Fig. 3.58a and Fig. 3.58b for the TTD. It is shown, that the convolution of the magnetic flux densities also occurs in the temporal

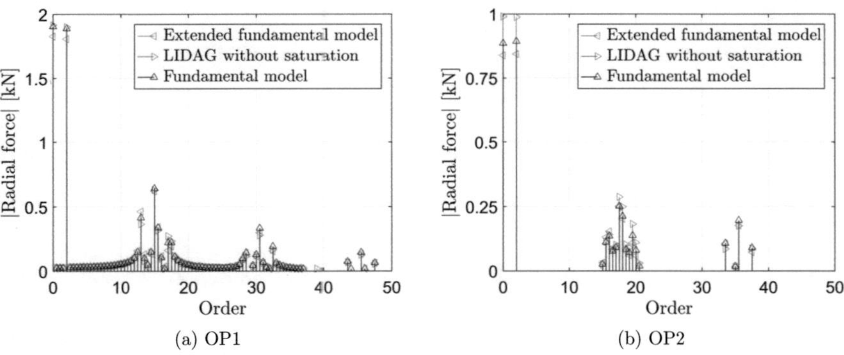

Figure 3.57: Temporal comparison of radial forces beneath one stator tooth for (a) OP1 and (b) OP2 in TFD

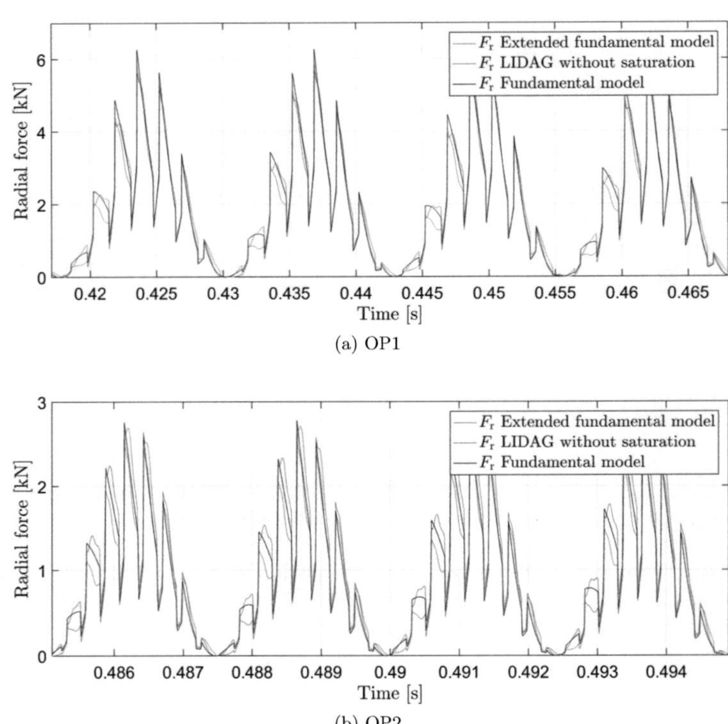

(a) OP1

(b) OP2

Figure 3.58: Temporal comparison of radial forces beneath one stator tooth for (a) OP1 and (b) OP2 in TTD

time domain. Differences are increased and decreased as well, due to the convolution of the frequency components.

3.3.1.2 Comparison between FEM and Integrated Electromagnetic Model

The comparison between the electromagnetic FEM model and the integrated electromagnetic model, especially the LIDAG includes the complete machine saturation behavior. The induction machine operation is shown for the given OPs from table 3.1 and is also fed with sinusoidal voltages.

Stator current

The saturated stator U-phase currents are shown for operating point 1 in Fig. 3.59a in the TTD and in Fig. 3.59b in the TFD. The position of the stator currents matches, but the magnitudes of the occurring orders show differences. The 1st temporal order of the LIDAG is slightly different to the FEM, caused by the saturation, which is rated higher, causing the higher 5th order. This changes the amplitudes of the slotting harmonics of 14.12th and 16.12 order as well as the 20.12th and 22.12th inducted by the slip relating rotor slotting $\mu = 17\text{th},19\text{th},\ldots$ order in addition to the saturation 5th and 7th order.

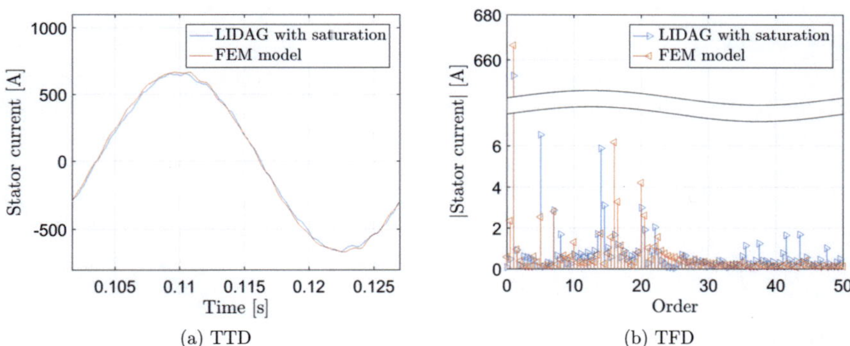

(a) TTD   (b) TFD

Figure 3.59: Stator phase currents for OP1 in (a) TTD and (b) TFD

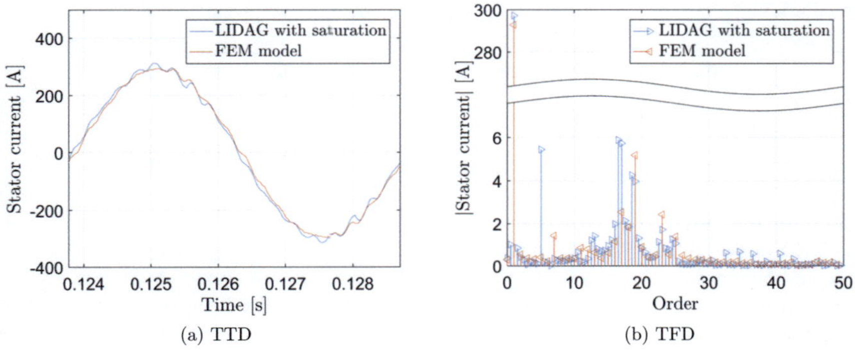

(a) TTD   (b) TFD

Figure 3.60: Stator phase currents for OP2 in (a) TTD and (b) TFD

Amplitudes of higher harmonics are attributed to this effect and to the slot width, that is given as constant factor, changing in simulation the slotting depth by saturation.

Fig. 3.60a and Fig. 3.60b show the simulation results for OP2 in the TTD and the TFD. The position of the stator currents are matching, but show the same behavior like OP1 in the frequency domain. High saturation causes a higher 5th, which leads in this case to a matching fundamental amplitude, but changes the ratio slotting harmonics of 16.72nd and 18.72nd as well as the 22.72nd and 24.72nd order. The ratio change is induced by the slip depending rotor slotting 17th and 19th order as well as the saturation 5th and 7th order. Amplitudes of higher harmonics are also attributed to saturation and slot width dependency.

Rotor current

The saturated rotor bar currents are shown for OP1 in the TTD in Fig. 3.61a and in the TFD in Fig. 3.61b. The position of the rotor current of the LIDAG is matching with the FEM results. The analysis in TFD shows, that the LIDAG assess the induced harmonics from stator side from $\nu = $ 5th,7th,... resulting in rotor current harmonic orders 30.61st and 32.61st, higher than the FEM simulation.

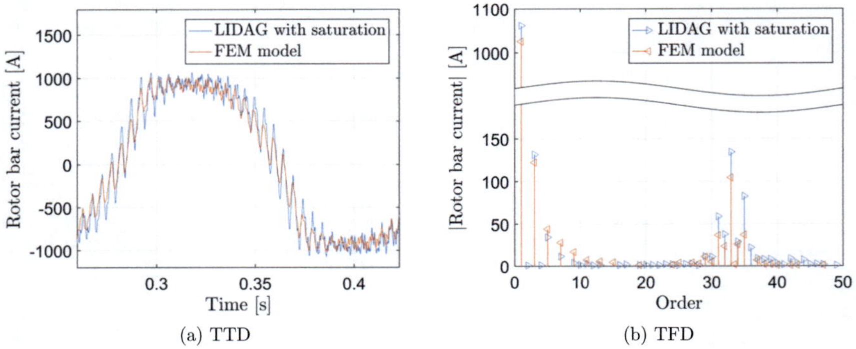

(a) TTD   (b) TFD

Figure 3.61: Rotor bar currents for OP1 in (a) TTD and (b) TFD

The rotor bar currents for operating point 2 are shown in the TTD in Fig. 3.62a and in the TFD in Fig. 3.62b. The position of the fundamental wave is matching and it shows a marginal difference in the amplitude. The slotting harmonics, represented by the 5th and 7th order show the same behavior. However, the orders of the higher harmonics, induced from the stator $\nu = $ 5th,7th,... behave differently. Using the slip dependent equation (3.12), the induced harmonics are calculated to the 383.29th and 381.29th matching with the LIDAG simulation. The deviation to the FEM simulation can not be explained. However, it is

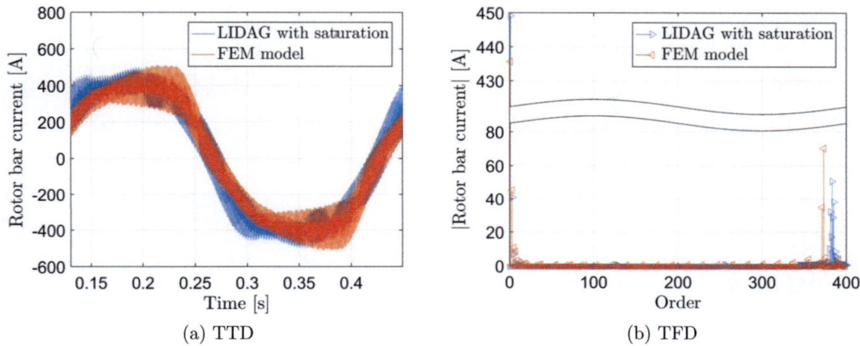

(a) TTD

(b) TFD

Figure 3.62: Rotor bar currents for OP2 in (a) TTD and (b) TFD

worth mentioning, that the FEM simulation could potentially have issues calculating very high temporal orders, caused by solver errors.

Electromagnetic torque

The electromagnetic torque including the saturation effects is shown in Fig. 3.63a and Fig. 3.63b for operating point 1. Figure 3.64a and Fig. 3.63b show the effects for operating point 2. The graphs disclose, that the fundamental amplitude matches, but the saturation harmonics of the 6th order is rated higher in the LIDAG than in the FEM model. Slotting harmonics 15.12th at OP1 and 17.72nd at OP2 and the superposition of saturation with slotting 21.12th in OP1 and 23.72nd in OP2 are rated lower. Harmonics, that are higher than the 50th order also have lower amplitudes in the LIDAG. This is also caused by the saturation and the slotting width, that can have a huge influence.

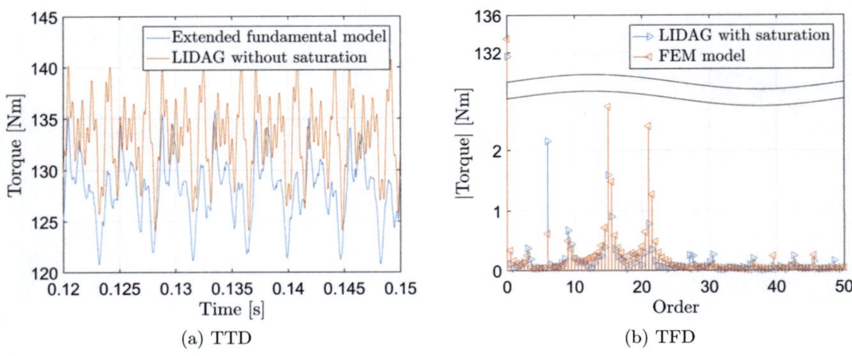

(a) TTD

(b) TFD

Figure 3.63: Electromagnetic torque for OP1 in (a) TTD and (b) TFD

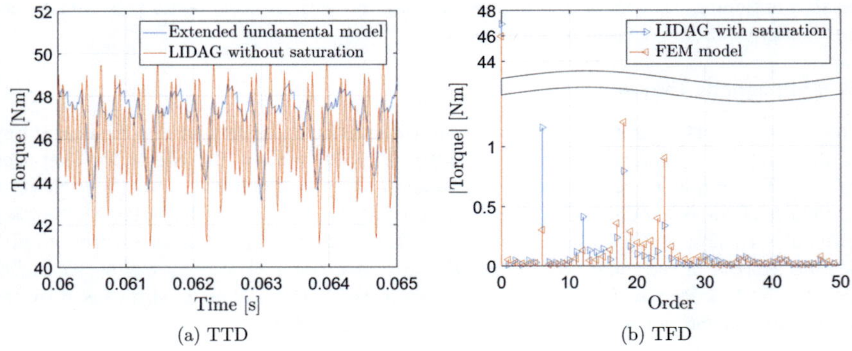

(a) TTD

(b) TFD

Figure 3.64: Electromagnetic torque for OP2 in (a) TTD and (b) TFD

Magneto motive force

The mmfs of the FEM simulation is synthesized using the spatial stator and rotor current distribution for each time step as well as the mechanical rotor movement, coupled with the winding functions of the sequential electromagnetic model. The spatial comparison of the mmfs are shown for OP1 and OP2 in the STD in Fig. 3.65a. Figure 3.65b exhibits the comparison for one time step and the temporal comparison of the mmfs beneath one stator tooth in the TTD and the TFD in Fig. 3.66a and in Fig. 3.66b. These Figures show, that there are only small differences occurring due to the harmonics content of the currents, but the positions are matching. Thus, deviations in the amplitudes of the summed mmf Θ are explained by the small differences in the amplitudes of the induced harmonics.

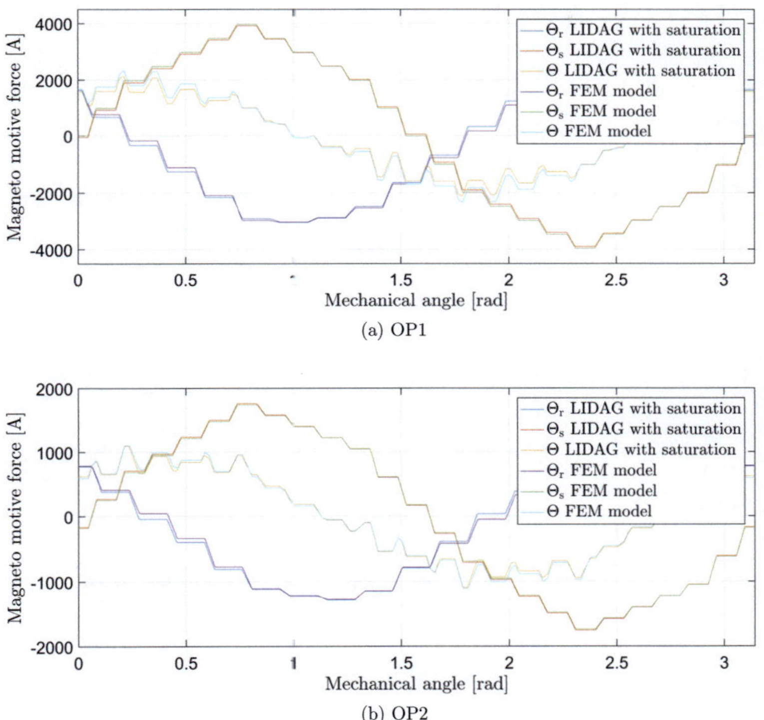

(a) OP1

(b) OP2

Figure 3.65: Spatial comparison cf mmfs at one time step for (a) OP1 and (b) OP2 in STD

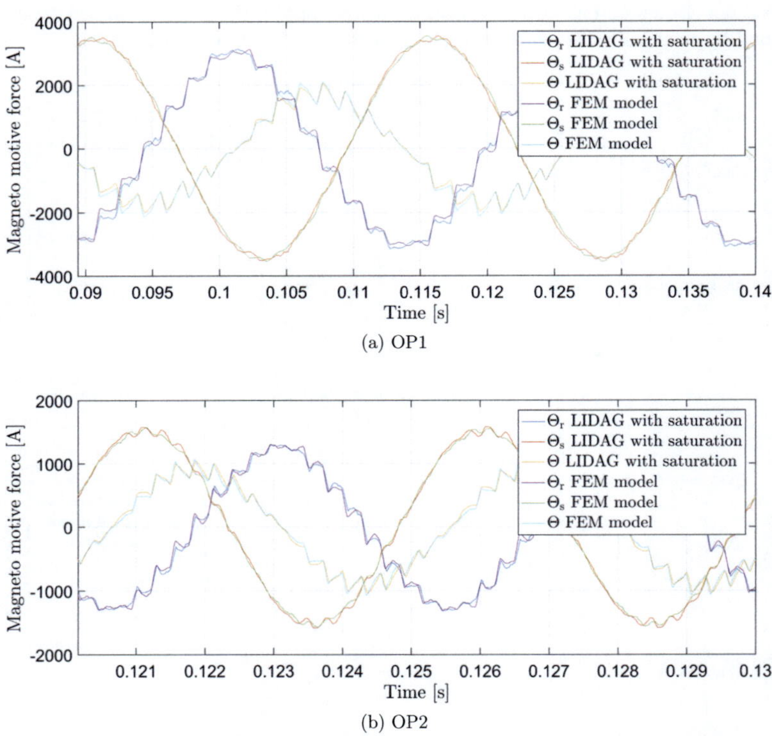

(a) OP1

(b) OP2

Figure 3.66: Spatial comparison of mmfs beneath one stator tooth for (a) OP1 and (b) OP2 in TTD

Magnetic flux density

The spatial comparison of the magnetic flux densities between the LIDAG and the FEM model for one time step is shown in the STD in Fig. 3.67a for OP1 and in Fig. 3.67b for OP2. Small differences occur due to the tooth saturation, the rectangular approach of the slotting as well as the change of the slot width relating to the operating point, which is included in the FEM simulation, but neglected in the LIDAG. These effects are also observable in the SFD in Fig. 3.68a and Fig. 3.68b. Especially the rotor slotting harmonics of $\mu = 35\text{th}, 37\text{th}$ are rated too high, because of the change of the slot width when the tooth saturates.

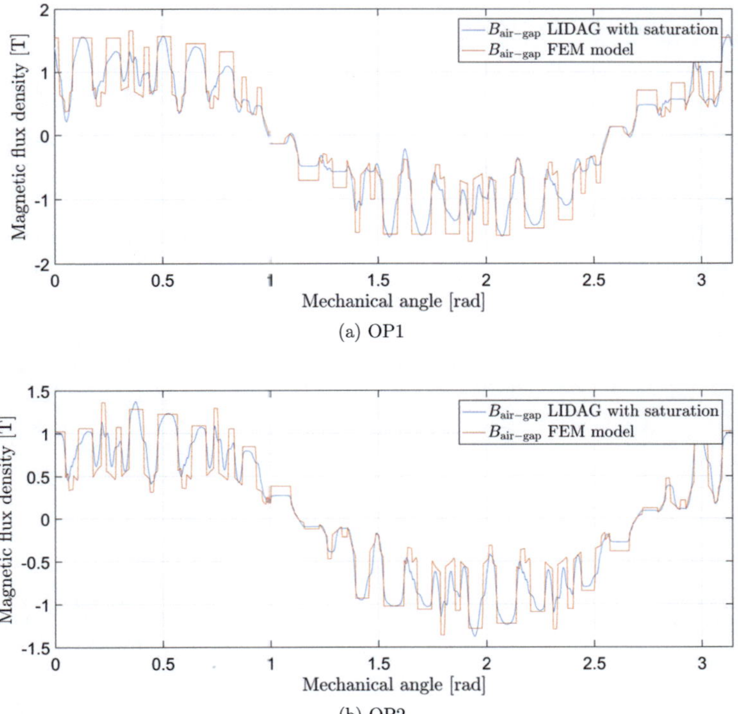

(a) OP1

(b) OP2

Figure 3.67: Spatial comparison of magnetic flux densities at one time step for (a) OP1 and (b) OP2 in STD

Figure 3.70a and Fig. 3.70b show the temporal comparison of the magnetic flux densities beneath one stator tooth for the given operating points in the TTD. Figure 3.69a and Fig.

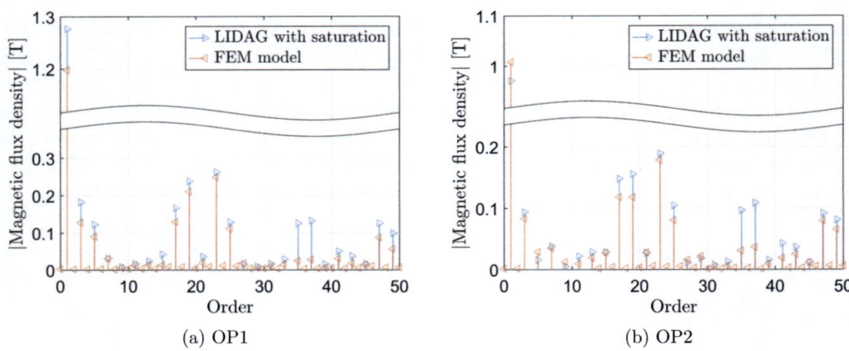

(a) OP1   (b) OP2

Figure 3.68: Spatial comparison of magnetic flux densities at one time step for (a) OP1 and (b) OP2 in SFD

3.69b illustrate the comparison in the TFD. These figures show a similar behavior to the spatial comparison.

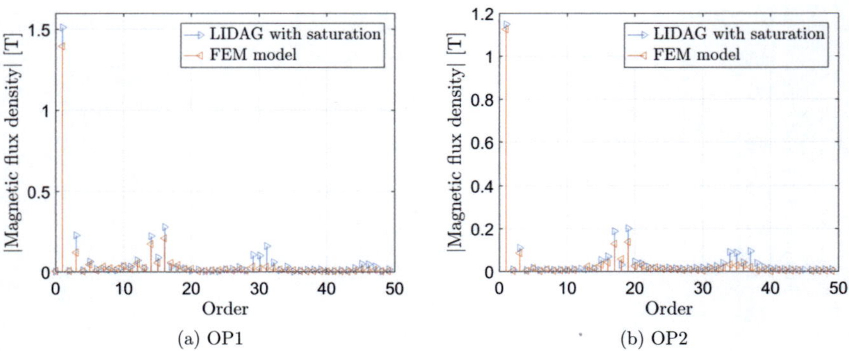

(a) OP1   (b) OP2

Figure 3.69: Temporal comparison of magnetic flux densities beneath one stator tooth for (a) OP1 and (b) OP2 in TFD

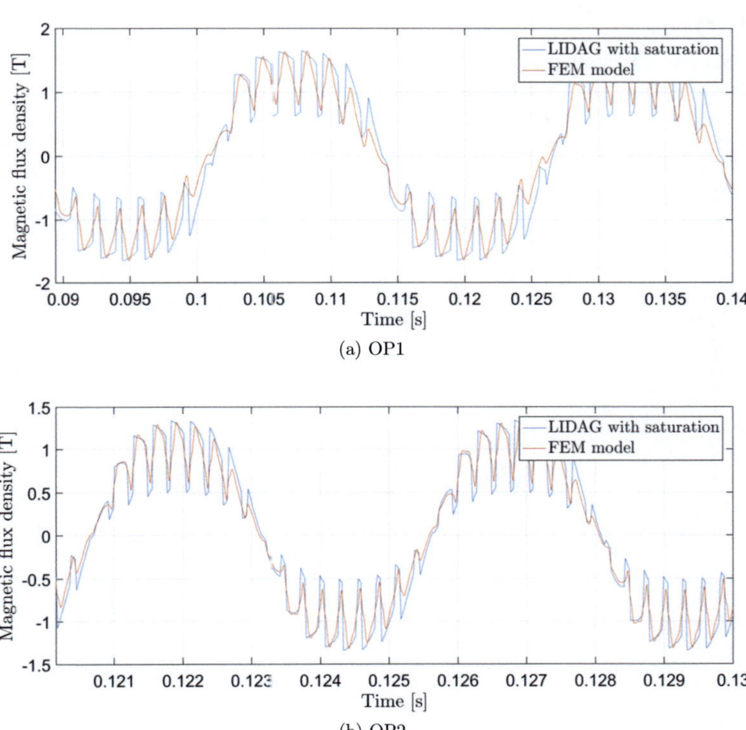

(a) OP1

(b) OP2

Figure 3.70: Temporal comparison of magnetic flux densities beneath one stator tooth for (a) OP1 and (b) OP2 in TTD

Radial force

The radial forces including saturation are shown at one time step for OP1 and OP2 in Fig. 3.71a and Fig. 3.71b in the STD as well as in Fig. 3.72a and Fig. 3.72b in the SFD . The spatial positions are matching, showing differences in saturation behavior and slot width. In the SFD the spatial orders are identical. However, for both OPs, the amplitudes of the harmonic distribution of the LIDAG are higher in comparison to the FEM. This effect occurs due to the convolution of the magnetic flux density with itself, where the superposition of the slotting harmonics with the other orders lead to higher amplitudes in the LIDAG.

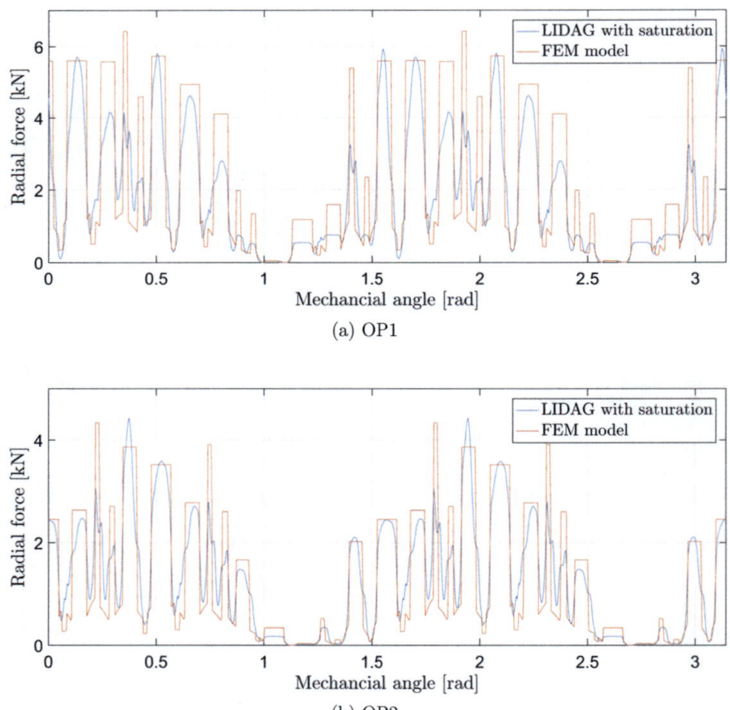

(a) OP1

(b) OP2

Figure 3.71: Spatial comparison of radial forces at one time step for (a) OP1 and (b) OP2 in STD

The temporal comparison of the radial forces is shown in Fig. 3.73a and Fig. 3.73b for the TTD and in Fig. 3.72a and Fig. 3.72b for the TFD. The positions of the radial forces are matching. The behavior of the harmonic distribution can be attributed to the similar causes due to the spatial comparison.

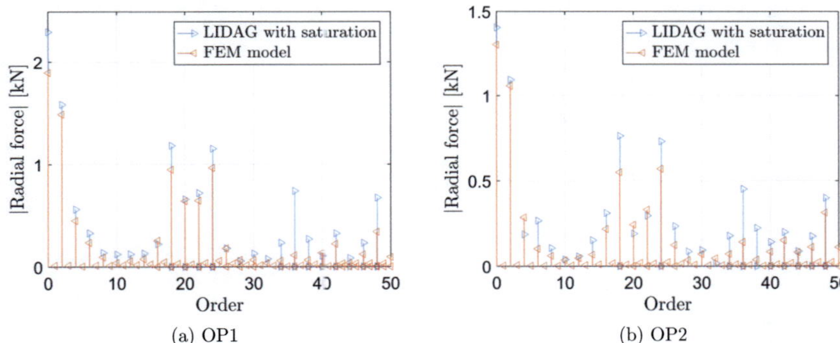

(a) OP1   (b) OP2

Figure 3.72: Spatial comparison of radial forces at one time step for (a) OP1 and (b) OP2 in SFD

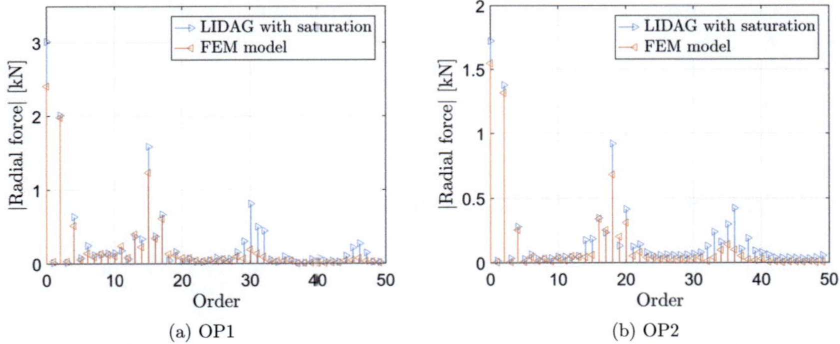

(a) OP1   (b) OP2

Figure 3.73: Temporal comparison of radial forces beneath one stator tooth for (a) OP1 and (b) OP2 in TFD

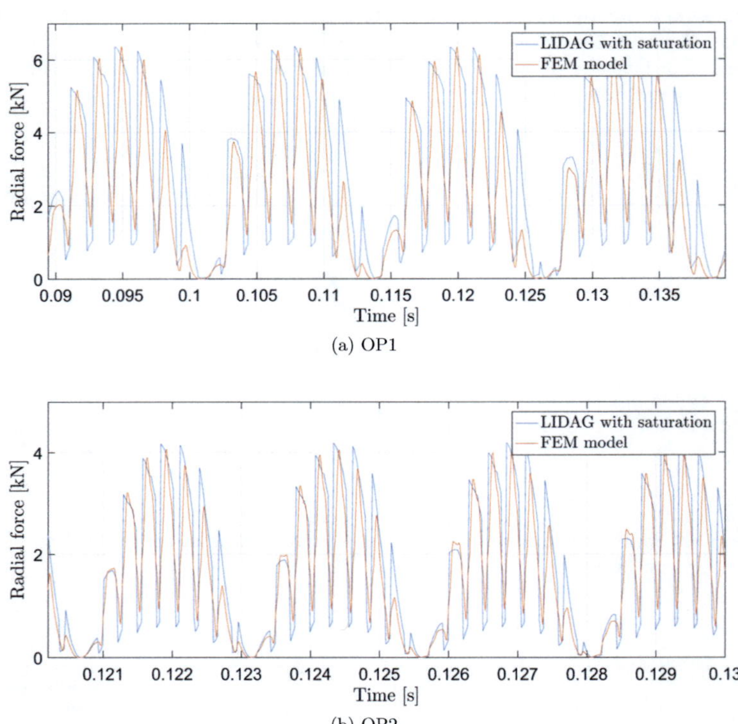

Figure 3.74: Temporal comparison of radial forces beneath one stator tooth for (a) OP1 and (b) OP2 in TTD

3.3.2 Validation with Measurements

The validation with test bench measurements is performed with a speed run-up from 0rpm to 18000rpm at 118Nm of torque. The induction machine is fed with an PWM-Inverter, operating with SVPWM at 10kHz and a 180V link voltage. In this validation, the stator currents and the mean torque are compared to each other. Other quantities like the magnetic flux density in the air-gap of the machine and the rotor bar current measurement for the induction machine cage rotor are not performable. However, the magnetic flux density and the resultant forces are validated with the surface velocity measurements after the introduction into the structural dynamic chapter 4, transferring the radial air-gap forces into the stator surface velocities.

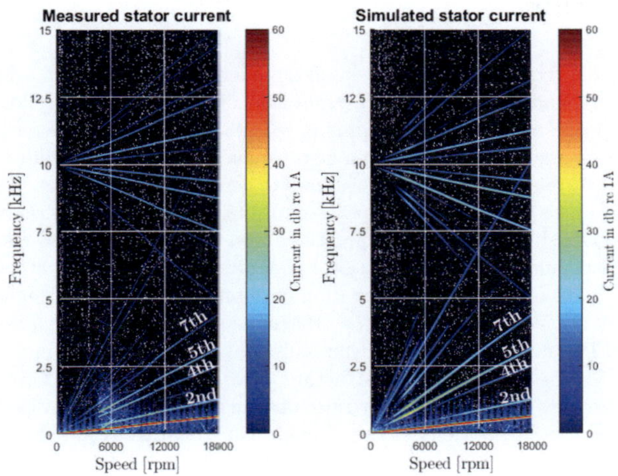

Figure 3.75: FT based spectrum of the stator current U-phase for (left) measured and (right) simulated speed run-up from 0 rpm to 18000rpm at 118Nm of desired torque

Figure 3.75 shows the spectrograms with its harmonic amplitudes plotted over frequency and rotor speed. The prominent bands can be seen in the spectrum, below is the fundamental order along with side bands from the voltage excitation at 2nd, 4th, 5th and 7th order as well as the saturation 5th and 7th order, the initial rotor slotting orders and the saturation-slotting interaction. Additionally, the PWM harmonics along with its bands emanate at around 10 kHz.

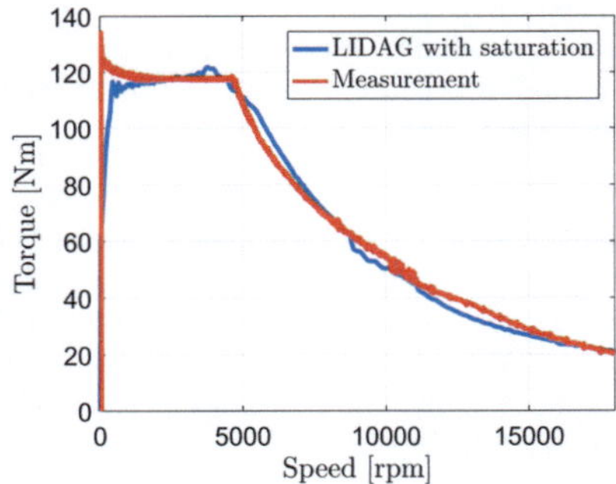

Figure 3.76: Comparison of torque for a run-up from 0 to 18000rpm at 118Nm of desired torque

Figure 3.76 shows the filtered torque over speed for a run-up at 118Nm of desired torque. The general trend conforms the saturation functions introduced in the LIDAG. Differences occur due to the parametrization of the LIDAG as well as the parametrization of the test bench observer and test bench controller settings.

3.4 Conclusion

This chapter introduced electromagnetic simulation models for induction machines. In order to achieve a short simulation times with transient validity and high accuracy, this work suggests the following modifications. Beginning with the current response for the sequential electromagnetic simulation method, harmonic rotor current circuits are added in parallel to the fundamental current calculation. This allows to generate the induction of spatial stator harmonics. This procedure leads to comparable results, but also entails disadvantages with regards to the phase shift, due to the indirect addition of inductances and resistances. The saturation behavior, inducing the saturation harmonics, is not included either. This is not performable for additional temporal saturation harmonics with the IRTF model structure. The current simulations with the sequential electromagnetic simulation can be linked to the STD and the SFD. In this case, the calculation of the electromagnetic induction machine behavior in the SFD is a refined model structure. On the one hand, the force model in the STD shows advantages regarding the saturation calculation. But entails disadvantages due to the discretized spatial positions in the air-gap. On the other hand, the force model in the SFD only allows simplified saturation functions, without the transformation back into the STD, but includes advantages especially in the minimization of numerical errors due to the steady spatial functions. Additionally the model allows to save the transformation of the STD into the SFD for the structural dynamic calculation in chapter 4. In order to create a model, which includes the spatial and temporal slotting and saturation effect in a content causality chain, the integrated electromagnetic simulation model is implemented, which is also a refined method in order to calculate the electromagnetic quantities of the induction machine. With this model structure, currents are directly calculated of spatial and temporal depending magnetic flux linkages and subsequently of the induction machines' magnetic flux density.

The proposed models are compared to and validated against each other, including the electromagnetic FEM simulations. Table G.1 gives an overview of the operating points. Hereby, the fundamental torque, stator and rotor currents are opposed for the simulation methods. The total harmonic distortion (THD) shows the harmonics content of the signals and gives a rough hint of the accuracy of the results. The deviation of the values are caused by the different saturation behavior, due to the induced harmonics. Additionally the variation of the slot width arises the penetration of the magnetic flux density into the slots. Thus, the results show minor differences, but still have a high accuracy.

To sum up the sequential and integrated electromagnetic simulation models, the saturation behavior, changing the entering magnetic flux density, can be improved for steady state results changing the winding and slotting functions depending on the saturation depth. The saturation also changes the leakage inductance of the stator and rotor, affected by the flux path through the induction machine. This affects the rectangular functions of stator and rotor winding function as well to a smoother winding function. Figure 3.77a shows the end winding structure of the induction machine, when changing its length from $\bar{c}d$ to $c'd'$

(Fig. 3.77b). With these improvements, the induced saturation harmonics into the stator and rotor bar currents can be performed with a higher accuracy.

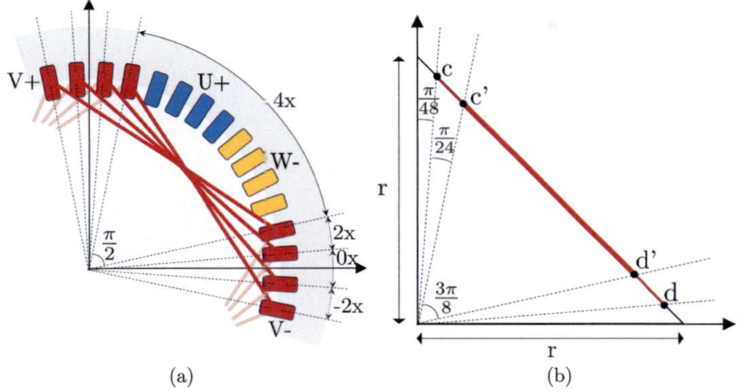

Figure 3.77: (a) Upper end winding connection of the stator u-phase for a 48 slot stator 2 pole pair machine and (b) Mean line segment for the 4 windings with effective turns strength of 4 for c'd' and 2 for cc' and d'd

4 Structural Dynamic Modeling

The structural dynamic model applies the electromagnetic forces from the air-gap of the induction machine over the stator structure to the outer stator surface. This creates vibrations on the surface and results in the induction machines' surface velocity. With this information, the perceived sound pressure level can be calculated, which is radiated into the environment. The introduced procedure is not only applicable to the induction machine but is also transferable to any type of electrical machine. For the structural dynamic simulations, spatial orders are called modes, assuming exciting force modes and structural modes relating to linear superposition [GWC06].

Referring to the simulation path from Fig. 1.5, this chapter relates to the second part of the structural synthesis. Applying the radial electromagnetic air-gap forces from chapter 3, the forces have to be first decomposed and transformed into the SFD and TFD in section 4.1. With this decomposition, the frequency components of the air-gap forces can be used. The air-gap force interaction with the inner stator surface, mainly interacting with the stator tooth, for the structural dynamic model, is outlined in section 4.2. Before using the structural dynamic model, precalculations are needed, using a modal, harmonic analysis, as well as the force response calculation of the stator structure, in section 4.3. Finally, the surface displacement synthesis is performed in section 4.4.

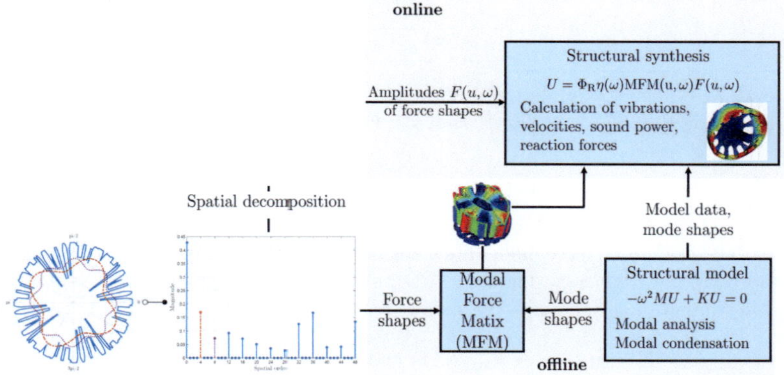

Figure 4.1: System integration of the structural dynamic model

4.1 Radial Air-gap Force Decomposition

In order to analyze the content of the radial force and to couple it with the structural dynamics of the stator, the radial force has to be decomposed spatially and temporally. This procedure can be done like a 2-D FFT, transforming the radial force, shown in (4.1), first spatially and then temporally, as well as vice verse [Boe13; Bes08; Sei92; Kot+17b; Boe+10; Kot+17a; VBG93; VBG92].

$$F_{\mathrm{r}}(\alpha,t) = \frac{B_{\mathrm{air-gap}}^2(\alpha,t)}{2\mu_0} A_{\mathrm{surf}}, \; F_{\mathrm{r}}(\alpha,t) \in \mathbb{R} \tag{4.1}$$

Spatial decomposition

After the calculation of the magnetic flux densities and subsequently the radial forces in the air-gap of the induction machine, as introduced in chapter 3, the radial force density is spatially decomposed for each time step, as shown in (4.2), where the coefficients $F_{\mathrm{r,cos}}(u,t) = \Re\{\underline{F}_{\mathrm{r}}(u,t)\}$ and $F_{\mathrm{r,sin}}(u,t) = \Im\{\underline{F}_{\mathrm{r}}(u,t)\}$ for the force mode u are varying with time.

$$F_{\mathrm{r}}(\alpha,t) = \sum_{u=0}^{U} F_{\mathrm{r,cos}}(u,t)\cos(u\alpha) + F_{\mathrm{r,sin}}(u,t)\sin(u\alpha), \; F_{\mathrm{r}}(\alpha,t) \in \mathbb{R} \tag{4.2}$$

The occurring spatial force modes for the symmetrical energized and healthy induction machine are defined in (4.3), with the machine pole pair number p. The spatial orders arise due to the mathematical convolution of the magnetic flux density with itself, derived in appendix F, in table F.3, freed from the machine pole pair number.

$$u = 2pk_{\mathrm{u}}, \text{ with } k_{\mathrm{u}} \in \mathbb{Z} \tag{4.3}$$

Figure 4.2 shows the spatial decomposition and the 1-D Fourier-approximation for one time step, according to (4.2). It is shown, that the force can be reconstructed with fewer spatial orders.

Temporal decomposition

The spatial decomposed force modes are transformed from the temporal time domain (TTD) into the temporal frequency domain (TFD), shown in (4.4), which allows to analyze the temporal content of the spatial force shape.

$$F_{\mathrm{r,cos}}(u,t) \circ\!\!-\!\!\bullet F_{\mathrm{r,cos}}(u,f) \text{ and } F_{\mathrm{r,sin}}(u,t) \circ\!\!-\!\!\bullet F_{\mathrm{r,sin}}(u,f) \tag{4.4}$$

The temporal signals of the radial force are shown in Fig. 4.3a and Fig. 4.3b in SFD before transforming them into the TFD. The temporal decomposition of force shape 0 and

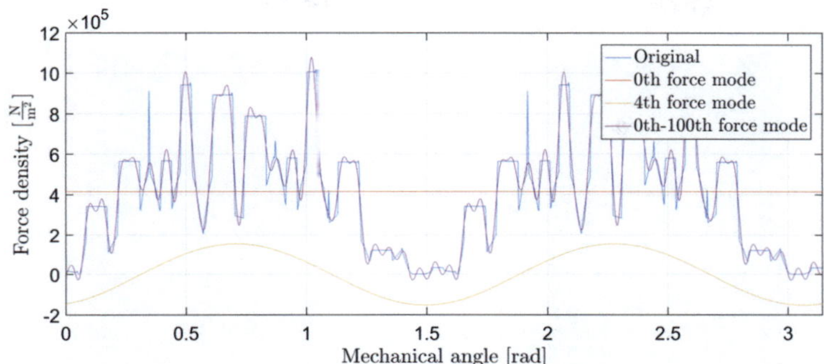

Figure 4.2: Spatial decomposition of the radial force

4 are diagrammed in Fig. 4.4a and Fig. 4.4b. The PWM harmonics at around 10kHz and the induction machine harmonics increasing their frequency from the coordinate origin are identified in chapter 3.2.2. It is shown, that the radial force modes have a linear scaling of their individual frequency content with rotation velocities. Frequencies, which are included in the spatial orders are defined by temporal components of the spatial orders of the electromagnetic components, shown in appendix F.

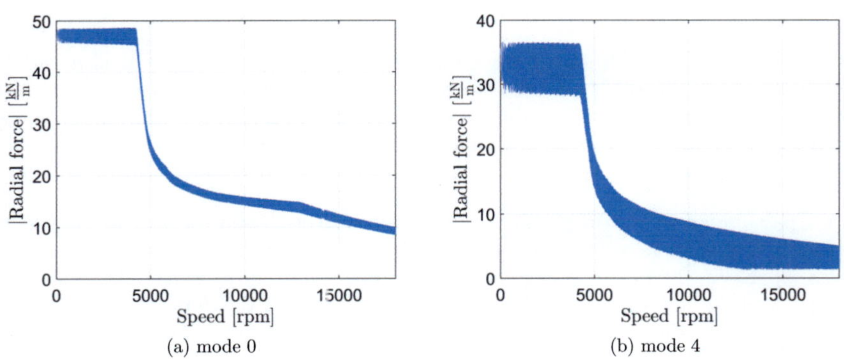

(a) mode 0

(b) mode 4

Figure 4.3: Magnitude of the radial force (a) mode 0 and (b) mode 4 in TTD

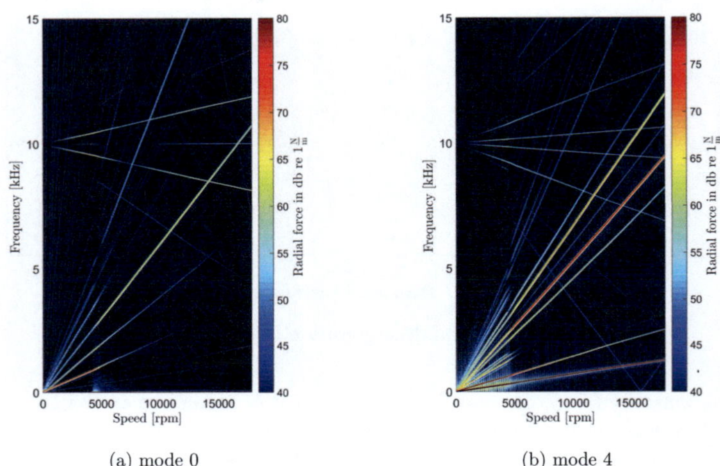

(a) mode 0　　　　　　　　　　　　　　(b) mode 4

Figure 4.4: Temporal decomposition of radial force (a) mode 0 and (b) mode 4

4.2 Stator Tooth Force

Before coupling the air-gap force with the structural synthesis to the electrical machine's structural model, the air-gap force needs to be applied on the inner stator surface. This step is important to reduce the creation of artifacts in the structural system and can change the influence of different force modes, changing the amplitudes and phase angles of spatial modes [Boe13; Boe+12; Boe+10; Gar+97].

In the past literature, the air-gap force was projected to the stator tooth based on various methods. One of these methods is the so-called, sine approach [Boe13; Boe+12; Boe+10]. In this approach, the spatial air-gap force distribution is directly applied to the mechanical dimension of the stator teeth, cutting out forces which arise in the slots of the stator. This procedure shows a significant disadvantage caused by cutting out and adding harmonic effects and neglecting the energy balance of the stator structure.
The most common approach is the constant tooth force projection [Boe13; Gar+97]. This approach leads to spatial aliasing caused by averaging the force in between of the middle of two neighbored slots and calculating one force value for each stator tooth (H.1).
Another tooth projection is the tsine approach, which is a refined procedure of the tooth force projection [Boe13; Boe+12; Boe+10]. This procedure mainly uses the idea of the radial force in between of the middle of two neighbored slots, but without averaging the

force to one tooth value. In this approach, the tooth forces are still distributed in space, reducing the aliasing without neglecting the balance of the created energy (H.1).

The continuous tooth force approach (tsine) is also performable in the SFD, using the inverse transformation matrix $H_{\text{tsine}}^{-1}(u)$ to project the radial forces $F_{\text{r}}(u)$ to the tooth force $F_{\text{tsine}}(u)$ for each time step [Boe13], as

$$F_{\text{tsine}}(u) = H_{\text{tsine}}^{-1}(u)F_{\text{r}}(u) \tag{4.5}$$

The transformation matrix $H_{\text{tsine}}(u)$ is defined as,

$$H_{\text{tsine}}(u) = \begin{cases} c_{\text{ratio}}\text{sinc}(\frac{\tau_{\text{tooth}}}{2\tau}u - \frac{u-k_tQ_s}{Q_s}) & \text{, where } u = u + k_tQ_s,\ k_t \in \mathbb{Z} \\ 0 & \text{, otherwise} \end{cases} . \tag{4.6}$$

In case of a given mesh of points on the inner surface of the electromagnetic simulation, like in the electromagnetic FEM, the most accurate way is to create an equally mesh in the structural dynamic simulation, like in the structural FEM [KZW16; Kot+16]. With this approach, the forces on the boundary layer of the inner stator surface, from the electromagnetic simulation, are directly applied to the boundary layer of the structural dynamic calculation. With regards to this work, the electromagnetic forces are calculated analytically without the application of any mesh on the inner stator surface [Kot+17a; Kot+17b]. For this application, the previous mentioned approach, the tsine approach, is considered in this work (4.5).

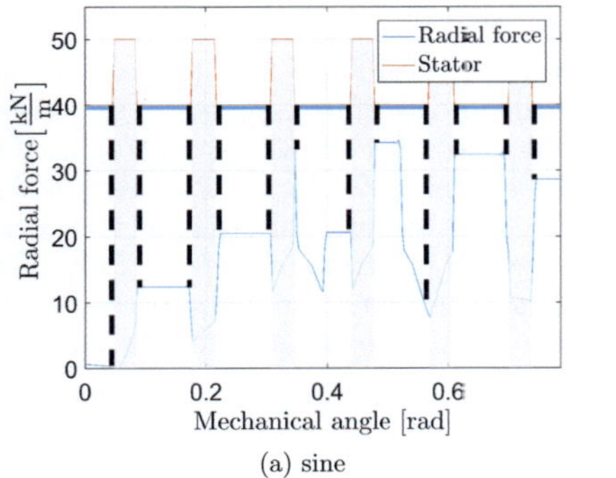

(a) sine

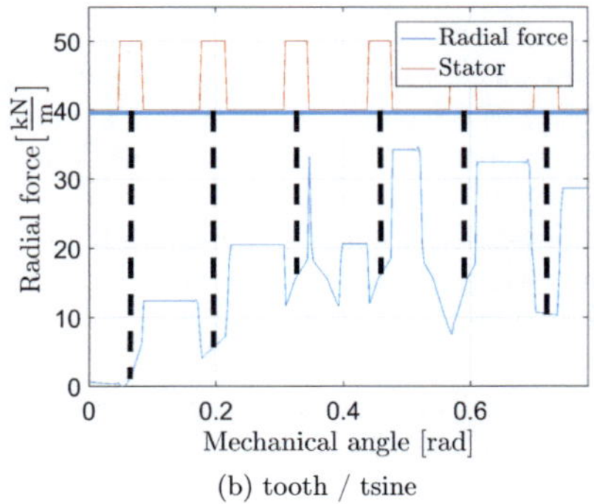

(b) tooth / tsine

Figure 4.5: Stator tooth force projection (a) sine and (b) tooth / tsine

4.3 Structural Response Analysis

The structural response analysis forms the basis for the surface displacement synthesis in section 4.4, where the stator tooth forces are coupled with the structural dynamic. In the past literature modal analysis in (App. I), the harmonic analysis (App. J) as well as the harmonic superposition (App. K) formed the basis in order to create the structural transfer function linking the tooth force to the surface vibration [GWC06; Le +09; Boe13; Dos88]. With these approaches, the surface displacement $u(\omega)$, including a maximum number of structural eigenforms M, with the transfer function $H_{\text{struct,m}}(\omega)$ can be written as,

$$u(\omega) = \sum_{\text{m}=0}^{M} H_{\text{struct,m}}(\omega) f_{\text{m}}(\omega). \tag{4.7}$$

The transfer function $H_{\text{struct,m}}(\omega)$ shows a dependency on mass m, the damping ξ and the angular eigenfrequency $\omega_{0,\text{m}}^2$, shown in (4.8).

$$H_{\text{struct,m}}(\omega) = \frac{1}{(\omega_{0,\text{m}}^2 + j\omega 2\xi_{\text{m}}\omega_{0,\text{m}} - \omega^2)m_{\text{m}}} \tag{4.8}$$

These approaches can be refined using (4.9), with the vector of modal response degree of freedom Φ_{R}, spectrum matrix $\eta(\omega)$, as well as the modal force matrix (MFM) [KZW16; Kot+17b; Kot+17a].

$$H_{\text{struct}}(\omega) = \underbrace{\begin{bmatrix} | & | & & | \\ \phi_{0,\text{R}} & \cdots & \phi_{\text{M,R}} \\ | & | & & | \end{bmatrix}}_{=:\Phi_{\text{R}}} \underbrace{\begin{bmatrix} \eta_0(\omega) & \cdots & 0 \\ \vdots & \ddots & \vdots \\ 0 & \cdots & \eta_{\text{M}}(\omega) \end{bmatrix}}_{=:\eta(\omega)} \underbrace{\begin{bmatrix} MFM(0,0) & \cdots & MFM(J,0) \\ \vdots & \ddots & \vdots \\ MFM(0,M) & \cdots & MFM(J,M) \end{bmatrix}}_{=:\text{MFM}} \tag{4.9}$$

The transfer function $H_{\text{struct}}(\omega)$ is still transferring the radial forces from the inner stator surface to the outer stator surface, where the modal response of freedom matrix Φ_{R} can be seen as spatial points on the surface. The spectrum matrix $\eta(\omega)$ is derived from the modal and harmonic analysis, shown in appendix I and J as well as the application of the modal superposition outlined in appendix K [GWC06; Le +09; Boe13]. Using (4.8) as individual components of the spectrum matrix $\eta(\omega)$, the normalized spectrum component of the m-th mode shape is rewritten, as shown in (4.10). Regarding to the norm of the structural mode shape, the mass is set constant to $m_{\text{m}} = 1$ for all $m = 1, \ldots, M$ in case of mass normalization, where the damping factor ξ_{m} is variable.

$$\eta_{\text{m}}(\omega) = \frac{1}{\omega_{0,\text{m}}^2 + j\omega 2\xi_{\text{m}}\omega_{0,\text{m}} - \omega^2} \tag{4.10}$$

With the normalization, the spectrum components have to be scaled for the transfer function. This is performed using the modal force matrix (MFM), with the m-th mode shape in its columns and the j-th force shape in its rows as,

$$\text{MFM} = \begin{bmatrix} MFM(0,0) & \cdots & MFM(J,0) \\ \vdots & \ddots & \vdots \\ MFM(0,M) & \cdots & MFM(J,M) \end{bmatrix}. \tag{4.11}$$

In order to use (4.9), the MFM has to be calculated using the modal and harmonic analysis [Kot+17a]. Refining the process from the force response calculation (App. L), the tooth force in the STD and TTD, as shown in (4.12), is decomposed with the 2-D FFT into SFD and TFD, shown in (4.13). With this decomposition, the harmonic components of any force excitation is identified, where each component is the complex amplitude of a specific spatially and temporally coupled frequency.

$$F_{\text{tsine}}(\alpha,t) = \begin{bmatrix} F_{\text{tsine}}(0,0) & \cdots & F_{\text{tsine}}(\alpha,0) \\ \vdots & \ddots & \vdots \\ F_{\text{tsine}}(0,t) & \cdots & F_{\text{tsine}}(\alpha,t) \end{bmatrix} \tag{4.12}$$

$$F_{\text{tsine}}(u,\omega) = \begin{bmatrix} F_{\text{tsine}}(0,0) & \cdots & F_{\text{tsine}}(J,0) \\ \vdots & \ddots & \vdots \\ F_{\text{tsine}}(0,M) & \cdots & F_{\text{tsine}}(J,M) \end{bmatrix} \tag{4.13}$$

With this information, the force response calculation is performed for each specific force shape and temporal frequency analogous to the force response calculation (App. L), where the normalized force $\hat{F}_{\text{tsine}}(u,\omega) = 1N$ is applied to the inner stator surface structure with either standing waves (L.3) or rotating waves (L.4). With this procedure, the MFM is filled with complex amplitudes for each spatial surface point Φ_{R}, scaling the spectrum matrix, which transfers the stator tooth force to the outer stator surface vibration.

4.4 Surface Displacement Synthesis

After the calculation of the structural response analysis, in section L, a surface displacement synthesis is performed. In the past literature, the radial forces $F(\omega)$ were scaled linearly using the superposition with a set of J forces shapes, shown in (4.14). The derived transfer functions from App. L were used by a linear scaling and superposition [Boe13; Roi09; vdGie11], as shown in (4.15). With (L.2), the surface displacement, including J force shapes, is expressed in (4.15).

$$F(\omega) = \sum_{j=0}^{J} \mathcal{F}_j \hat{F}_j(\omega) \tag{4.14} \qquad u(\omega) = \sum_{j=0}^{J} H_{u_{\text{norm,j}}}(\omega)\hat{F}_j(\omega) \tag{4.15}$$

Similar to this procedure, the superposition of a set of J tooth force shapes is applied in order to calculate the stator surface displacement, acceleration or velocity. Refining (4.15), the surface displacement synthesis is defined, using the transfer function (4.9) shown in 4.16. This expression is valid for the entire frequency range, at every time step or working point of any considered operation cycle [KZW16; Kot+17b; Kot+17a].

$$u(\alpha,\omega) = H_{\text{struct}}(\omega)\underline{\hat{F}}_{\text{tsine}}(u,\omega) = \Phi_{\text{R}}\ \eta(\omega)\ \text{MFM}\ \underline{\hat{F}}_{\text{tsine}}(u,\omega) \qquad (4.16)$$

In order to analyze the induction machines' surface behavior, the surface velocity $v(\omega)$ for a force excitation $\underline{\hat{F}}_{\text{tsine}}(u,\omega)$ is calculated as shown in (4.17), whereby the surface acceleration is analogously defined in (4.18). This synthesis leads to the nodal velocity of each node of the induction machines' surface and is called the operational deflection. Especially the position dependent normal surface velocities, using the exclusively normal components of the surface velocity of the surface positions, are important for the analysis and the sound power calculation. Using the normal surface velocity also enables the use of the mean-square surface velocity, which shows a squared proportionality to the radiated sound power [Boe13].

$$v(\alpha,\omega) = \text{j}\omega H_{\text{struct}}(\omega)\underline{\hat{F}}_{\text{tsine}}(u,\omega) \qquad (4.17) \qquad a(\alpha,\omega) = -\omega^2 H_{\text{struct}}(\omega)\underline{\hat{F}}_{\text{tsine}}(u,\omega) \qquad (4.18)$$

4.5 Validation and Comparison

This section compares of the tooth force projection methods and the calculation methods of the structural TFs and shows examples for the occurring force shapes and structural shapes as well as their relationship to each other. Eventually, the surface velocity measurement and the simulation for speed run-ups from 0 to 18000rpm at 118Nm torque will be validated.

Before starting with the comparison and validation, three different types of stator models can be assessed, shown in Fig. 4.6. First, the complete stator model, which was applied in the structural FEM analysis will be detailed in the following. The model discloses the complete stator behavior with the highest computation effort and is usually used for visualizing the operational deflections applying at critical working points for the entire stator structure. The FEM data can be reduced, only considering the surface. The results can then be applied to compute the emitted sound power over the entire housing surface including the surface mesh for the stator teeth in order to couple the force shapes to the structural mode shapes according to the tooth force approach. The stator teeth models, which are shown and compared in section 4.5.2, are applied by a simple cylinder, and its analytical transfer functions (TFs) which has the lowest accuracy and the least computation efforts. This model includes the same computation efforts like the structural FEM. However, the TFs can be reused and the computation time can thus be reduced.

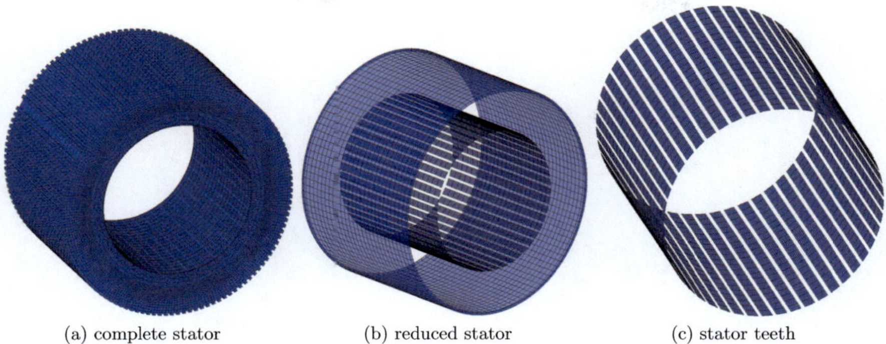

(a) complete stator   (b) reduced stator   (c) stator teeth

Figure 4.6: Comparison of (a) complete stator, (b) reduced stator and (c) stator teeth model.

4.5.1 Comparison of Tooth Force Simulations

It is important to discuss the different approaches for projecting the air-gap forces on the stator teeth. The calculation methods, shown in section 4.2, influence the amplitudes, the phase shift and the harmonics content of the radial force on the tooth. It is also important to implement the used method for the structural pre-calculation, to ensure the consistency of structural model.

Comparing the three methods, in Fig. 4.7 to each other, the sine and tsine approach mainly have sinusoidal functions, where the tooth projection is similar to a rectangular function. The sine and the tsine approach have small deviations in the phase shift and the amplitude, where the tooth force approach averages the harmonics content beneath two neighbored slots. This averaging decreases the amplitude and distorts the harmonics content beneath the stator teeth.

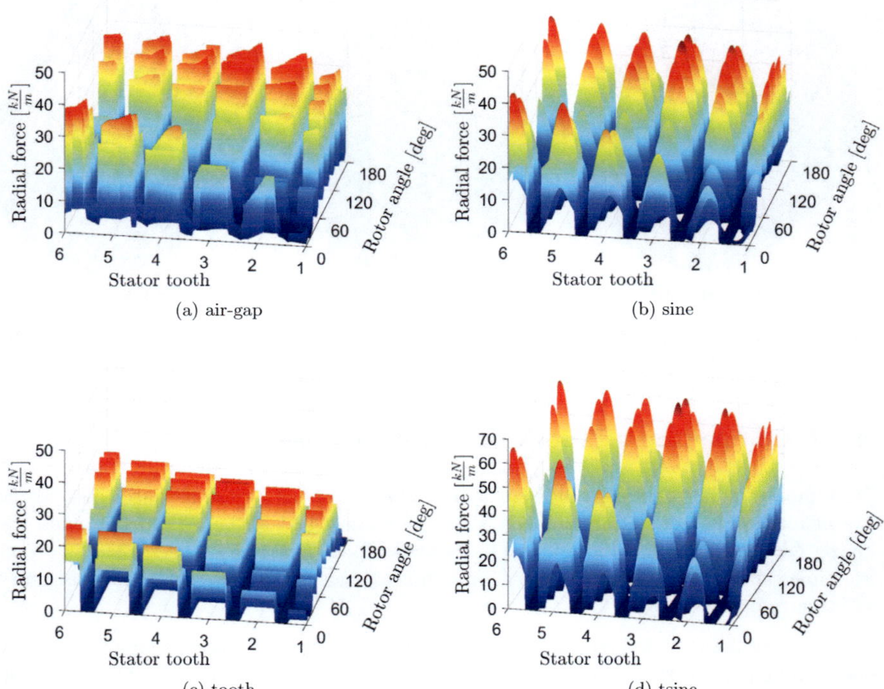

(a) air-gap

(b) sine

(c) tooth

(d) tsine

Figure 4.7: Comparison of radial forces for (a) air-gap, (b) sine, (c) tooth and (d) tsine approach.

4.5.2 Comparison of Structural Transfer Functions (TFs)

The structural transfer functions (TFs) describe the structural dynamics and reduce the complexity of the system. Using a simple infinite cylinder, the TF of the structure can be easily described and calculated with a low-pass filter of second order. In this case, each spatial force shape has one dominant peak, called eigen- or resonance frequency, shown in Fig. 4.8. In reality, the stator includes rows of rips as well as stator teeth. Additionally it includes a finite length, changing the TFs from the ideal analytical behavior or the infinite cylinder to a TF with numerous peaks. The cylindrical behavior with one dominant peak, calculated with the structural FEM is still displayed. As it is shown in Fig. 4.8, the

dominant behavior of the TFs for the 0th, 4th and 8th spatial force order are matching in the resonance frequency, but are dependent on the local position on the surface of the induction machine. Furthermore the amplitude changes, which increases the deviation to the analytical result. For spatial orders of 12 or higher, the resonance frequency and amplitude drastically deviate depending on the local position on the surface. Depending on the complexity of the machines' stator, the accuracy for the simple stator teeth model is only ensured for low spatial orders.

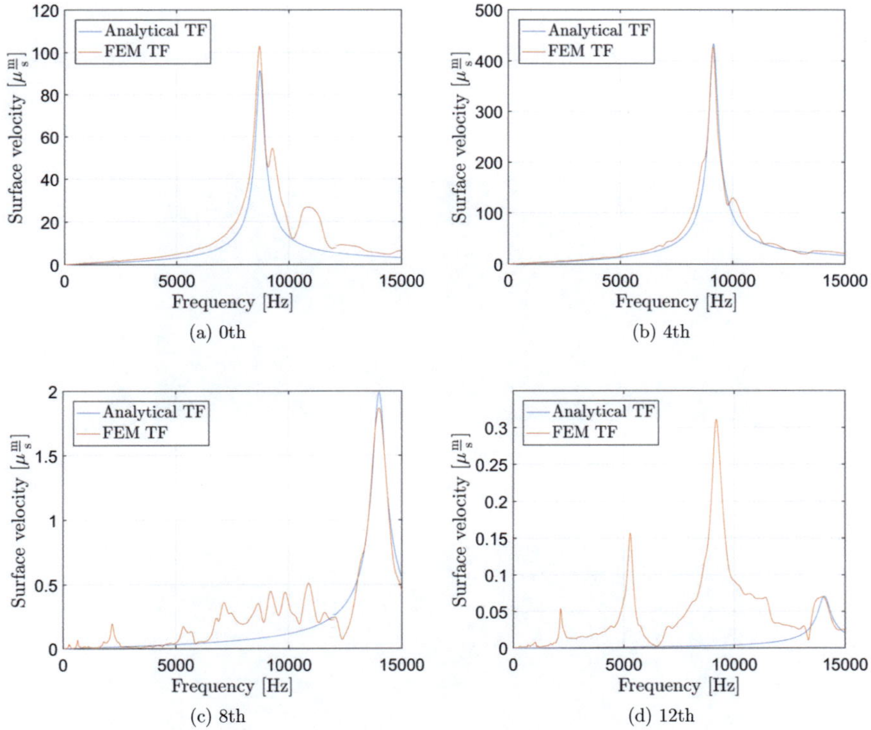

Figure 4.8: Comparison of TFs for the (a) 0th, (b) 4th, (c) 8th and (d) 12th force shape j.

4.5.3 Resultant Structure and Force Shapes

The resultant spatial force shapes of the 0th, 4th, and 8th order are shown in Fig. 4.9 for one time step, applying the harmonic force shape distribution to the stator teeth mesh of

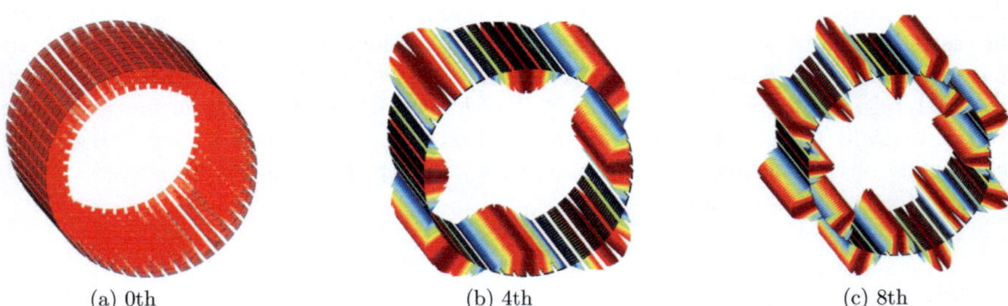

(a) 0th   (b) 4th   (c) 8th

Figure 4.9: Resultant (a) 0th, (b) 4th and (c) 8th force shape at one time step.

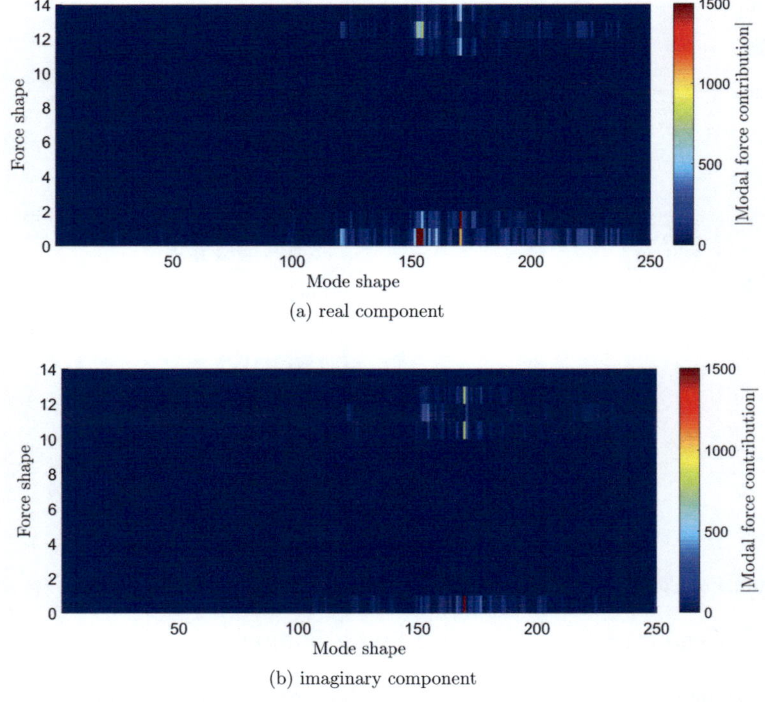

(a) real component

(b) imaginary component

Figure 4.10: Modal force matrix (MFM) for (a) real and (b) imaginary component

the structural FEM model. For the force response calculation, these forces are put in a structural FEM simulation, that changes their temporal swinging frequency and calculates the structural response for each temporal frequency. With this procedure, different structural shapes are excited, shown in Fig. 4.11. Examples for structural modes are, in this case, breathing shapes, breathing and bending shapes, bending shapes as well as axial- and radial-contraction shapes. Relating to the spatial force excitation and its frequency, many different structural modes are excited with different amplitudes, shown in Fig. 4.10. Additionally they are described as the MFM in section 4.3. The TFs include their different peaks and are created by the different temporal frequencies of the spatial force shapes from Fig. 4.8.

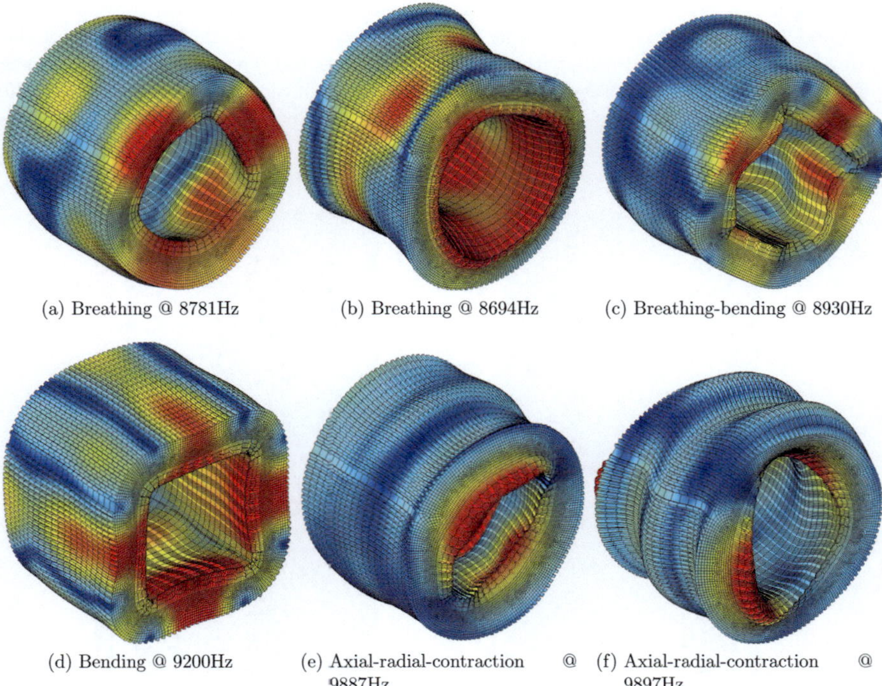

(a) Breathing @ 8781Hz   (b) Breathing @ 8694Hz   (c) Breathing-bending @ 8930Hz

(d) Bending @ 9200Hz   (e) Axial-radial-contraction @ 9887Hz   (f) Axial-radial-contraction @ 9897Hz

Figure 4.11: Resultant (a) breathing @ 8781Hz , (b) breathing @ 8694Hz, (c) breathing-bending @ 8930Hz, (d) bending @ 9200Hz, (e) axial-radial-contraction @ 9887Hz and (f) axial-radial-contraction @ 9897Hz structural mode shape.

4.5.4 Validation with Surface Velocity Measurements

The validation of the surface velocity is performed with speed run-ups from 0 to 18000rpm at 118Nm, shown in Fig. 4.12. Temporal machine orders from the convolution of the fundamental and the first order rotor slotting harmonics of the 17th and 19th create the $\sim$32nd, $\sim$36th as well as the $\sim$40th order. The machine orders from the convolution of the 3rd and 5th saturation harmonics with the first order rotor slotting harmonics of the 17th and 19th order create the $\sim$28th and $\sim$44th as well as the $\sim$24th and $\sim$48th order. In addition, they represent the dominant orders in the higher speed range of the investigated induction machine. Other important machine orders are created due to the convolution of the fundamental as well as the saturation harmonics slotting harmonics with the second order of slotting harmonics 35th and 37th, creating the orders around $\sim$72nd, in the speed range around 7000rpm and the machine orders due to the convolution of the first order rotor slotting harmonics 17th and 19th with itself, visible in the speed range of 3000rpm. Influences from the non-sinusoidal voltage excitation, caused by the inverter switching are also highly visible around the frequency bandwidth of 10kHz, created by the multiples of the switching and machine frequencies.

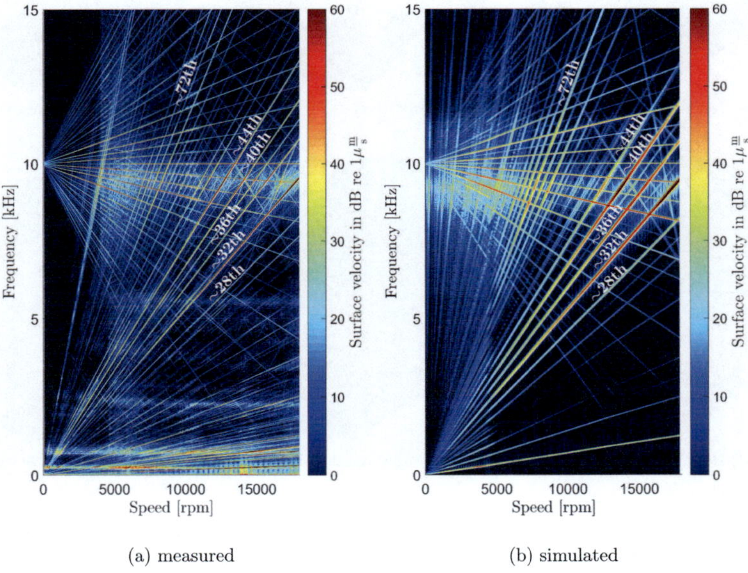

(a) measured   (b) simulated

Figure 4.12: Validation of (a) measured and (b) simulated FT based surface velocity spectrum.

In order to compare the spectrograms, the simulation shows well matching results to the measurement in the occurring orders and amplitudes. The orders deviate due to discretization errors, the resolution of the FT based spectrum, saturation and temperature differences as well as the rotor and bearing eccentricities.

The surface velocity levels are diagrammed in Fig. 4.13, where the surface velocity FT-based spectrum is averaged for each time step, or rather induction machine speed. This representation helps to verify the amplitudes of the harmonics content in the surface velocity. As shown in Fig. 4.13, the velocity levels of the measurement and simulation are in the same dimension and show the same behavior in principle. Some peaks in the measurement are not mirrored by the simulation, due to the already mentioned causes of saturation and temperature differences, discretization errors and deviations in the resolution of the FT based spectrum. In addition, peaks can occur due to imprecision in the manufacturing process, which can cause additional resonance frequencies as well as additional harmonic orders in the radial magnetic flux density in the air-gap of the induction machine. In general, the surface velocity level of the simulation follows the characteristics of the measurement.

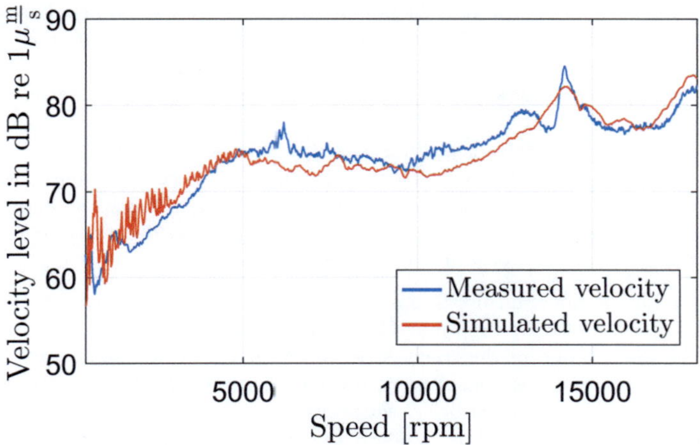

Figure 4.13: Surface velocity level for measured and simulated speed run-ups

4.6 Conclusion

At the beginning of this chapter, the spatial and temporal decomposition of the radial air-gap forces, from chapter 3, is proposed in section 4.1. This procedure allows to create steady functions from discretized functions and improves the way of analyzing the radial forces, exciting the structural dynamics model. As outlined in chapter 3, the radial forces can also be calculated directly in the SFD and only have to be transformed from the TTD into the TFD, which reduces the discretization error and the computation effort.
After the radial air-gap force decomposition, the different methods of tooth force projections are proposed in section 4.2. These methods show a high importance for the complete structural dynamics model and are thus clarified before starting with the analysis in section 4.3. These procedures have a huge influence on the tooth force excitation, especially on the amplitudes and the phase shift of the harmonics change with the proposed methods.

With the structural response analysis in section 4.3, based on the modal and harmonic analysis, the structural characteristics of the stator are introduced using TFs. Thus, this subsection defines the calculation of TFs, analyzing different structural shapes and their behavior. Additionally section 4.3 introduces the modal force matrix (MFM). The application of the modal force matrix (MFM) with the spectrum matrix is based on the principle of modal superposition, which allows the linear interaction of structural modes and force shapes. This new principle allows a much more detailed perspective and a better physical correlation.
In order to calculate the surface displacement, acceleration and velocity of the induction machine, the position depend normal velocity is proposed. Depending on the significance of the results, the surface displacement synthesis can be extended using the mean-square surface velocity for the sound power calculation and the position depend normal surface velocity for the position sound pressure perception position.

After the theoretical sections and discussions, the different tooth force methods are compared to each other. Furthermore the structural TFs for the stator models are opposed in section 4.5. With the presentation of the resultant structure and force shapes as well as their relationship, the validity is clarified. Finally, this section illustrates the comparison of the measured and simulated surface velocity, where the occurring orders relate to the radial air-gap force. As already described in chapter 3, the radial force is the convolution of radial magnetic flux density in the air-gap of the induction machine with itself and is damped by the structural dynamics. Thus, the simulation results relate to the electromagnetic and structural dynamics calculation and mostly show matching results. The deviations based on saturation and temperature differences, discretization errors and deviations in the resolution of the FT based spectrum as well as imprecisions in the manufacturing process and the mechanical borne harmonics due to bearings and air turbulence in the electrical machine.

5 Operating Point Adaption

The operating points (OPs) of the induction machine are often optimized for the lowest machine or drive losses. These points especially show an advantage in the thermal behavior of the drive system. However, the harmonic stimulation of the drive system is not taken into account. In order to consider the harmonics, which lead to vibrations on the surface of the electrical drive and audible noises, a new induction machines' operating strategy, for noise, vibration and harshness, is defined (Fig. 5.1). The operating points for one given torque are operable for different induction machine slip frequencies, changing the stator d-q-axis current operating strategy. The electromagnetic force excitation behavior is influenced, in order to subsequently reduce noises. First the operating strategies are introduced and the concept of changing the machines' operating strategy is explained. Finally, the measurement on the surface of the electrical drive is diagrammed [Bes08; BHK15; Bis+17; Bis+16; Kot+17b; Kot+17a; Kot+16], as well as the side effects of changing the operating strategy are analyzed.

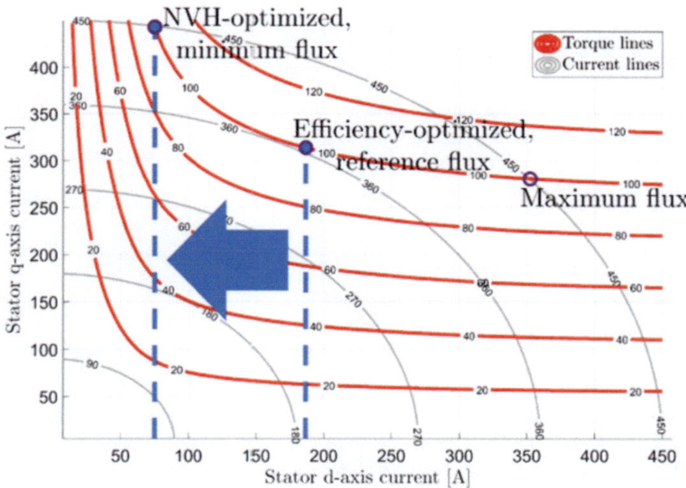

Figure 5.1: Schematic diagram of noise-optimized and efficiency-optimized operating points.

5.1 Concept for Operating Point Adaption

This section imparts the general understanding of the electromagnetic behavior of the electrical machine and subsequently to the electromagnetic force excitation, causing the noise emissions. Assuming that the radial electromagnetic force excitation F_r mainly depens on the radial component of the magnetic flux density $B_{air-gap}$, over a given surface area A_{surf} and the vacuum permeability μ_0, can be defined (5.1). The magnetic flux density $B_{air-gap}$ including the stator and rotor mmfs, Θ_s and Θ_r, as well as the magnetic permeance $\Lambda_{air-gap}$ are calculated as shown in (5.2), referring to chapter 3 [Bes08; Bis+17; Bis+16].

$$F_r(\alpha,t) = \frac{B_{air-gap}^2(\alpha,t)}{2\mu_0} A_{surf} \tag{5.1}$$

$$B_{air-gap}(\alpha,t) = \underbrace{(\Theta_s(\alpha,t) + \Theta_r(\alpha,t))}_{\Theta(\alpha,t)} \Lambda_{air-gap}(\alpha,t) \tag{5.2}$$

Considering the slotting of the stator and rotor, as well as the saturation in the magnetic permeance $\Lambda_{air-gap}$, both the mmfs of the stator and rotor, Θ_s and Θ_r, and their spatial position are the focused parts for changing the operating points and subsequently the noise emissions. In Fig. 5.2 - 5.3b the magneto motive forces are calculated for three different d-q-axis currents for a torque of 60Nm, as shown in the d-q-diagram in Fig. 5.1. The results reveal, that the position and the amplitude of the mmfs depend on the slip of the induction machine.

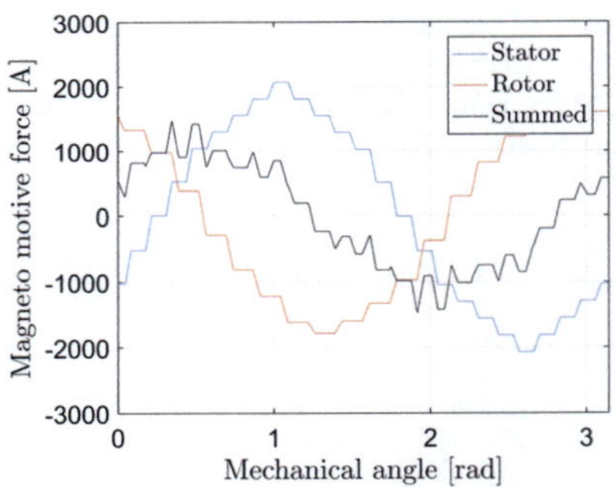

Figure 5.2: Magneto motive force (mmf) for reference stator d-axis current at 60Nm of torque.

The schematics for the fundamental electromagnetic are shown in Fig. 5.4a - Fig. 5.4c, where

$$\vec{i}_s(t) \propto \Theta_s(\alpha,t),\ \vec{i}_R(t) \propto \Theta_r(\alpha,t),\ \vec{i}_M(t) \propto \Theta(\alpha,t) \text{ and } \vec{\psi}_M(t) \propto B_{air-gap}(\alpha,t). \tag{5.3}$$

Decreasing the stator d-axis current decreases the flux and the magnetic flux density. This behavior is similar to the field weakening area of the machine operating range and increases the induction machine slip frequency, in order to hold the desired electromagnetic torque. Increasing the slip also increases the induced voltage into the rotor bars and also the rotor bar currents. As the rotor bar currents produce a mmf in order to reduce its origin, a field is built from the rotor side in order to damp the stator field. Increasing the induction,

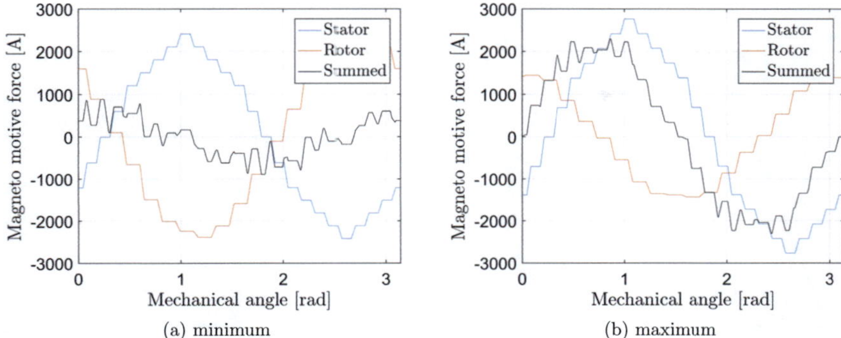

(a) minimum (b) maximum

Figure 5.3: Magneto motive forces (mmfs) for (a) minimum and (b) maximum stator d-axis current at 60Nm of torque.

decreases the spatial position difference between the stator and rotor mmf in comparison to the referenced system in Fig. 5.4a. This reduces the summed mmf, as well as the resultant magnetic flux density (Fig. 5.4b). Decreasing the slip, decreases the induction into the rotor bars. This decreases the occurring rotor bar currents and also decreases the amplitude of the rotor mmf. This results into a smaller opposing field from the rotor side. Due to this behavior, the rotor mmf damps the system less than the reference operating in Fig. 5.4a. The summed mmf, as well as the magnetic flux density in Fig. 5.4c increase. The angle between the stator and rotor mmf φ related to the stator and rotor currents $\vec{i}_s$ and $\vec{i}_r$ can be defined as,

$$\varphi(I_{d,min}) \leq \varphi(I_{d,ref}) \leq \varphi(I_{d,max}). \tag{5.4}$$

The harmonics content of each operating set at a torque of 60Nm is shown for simulated speed run-ups in the Campbell diagrams in Fig. 5.5a - 5.5c. These figures show the temporal frequency decomposition of the radial forces over the mechanical machine rotor speed. Since the voltage is produced by a PWM inverter, the voltage supply is non-sinusoidal and the same for each operating system. It is shown, that the d-q-axis operating point and subsequently the magneto-motive force, as well as the saturation of the magnetic permeance have a major influence on the electromagnetic force excitation, as explained in the following.

Keeping Fig. 5.5a as a reference, the operation in the minimum d-axis stator current in Fig. 5.5b shows less harmonics from the non-sinusoidal voltage excitation than the reference operating points. Due to the small spatial angle difference of the stator and rotor mmf, the rotor is able to damp more harmonics than in the reference system. The small angle and the increased rotor current reduces the fundamental amplitude of 2nd order as well as the

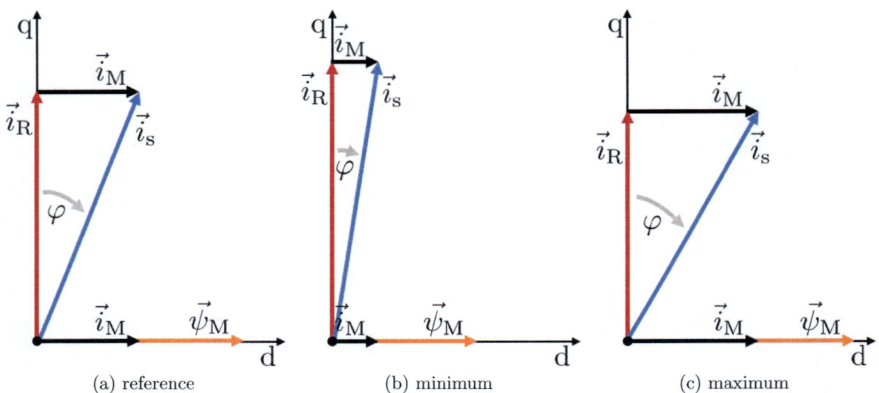

(a) reference   (b) minimum   (c) maximum

Figure 5.4: Schematic diagrams for (a) reference, (b) minimum and (c) maximum stator d-axis current at 60Nm of torque.

saturation harmonics of the 6th and 12th order. The mechanical harmonics, like the rotor slotting ~18th and ~36th order are more prominent, due to the 36 rotor bars and machine pole pair number two. The ~24th and ~30th harmonic order, caused by the interaction of saturation and slotting are reduced, due to a lower saturation, referring to section 3.2.2 [Bis+17; Bis+16; MVP08].

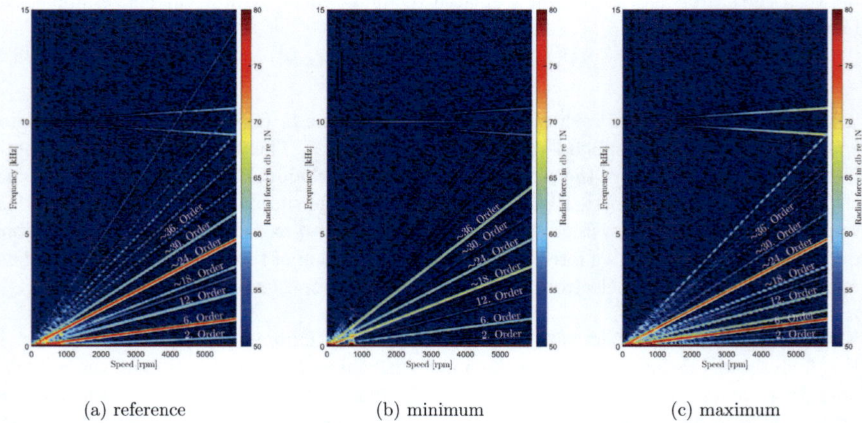

(a) reference   (b) minimum   (c) maximum

Figure 5.5: Radial force spectrum for (a) reference, (b) minimum and (c) maximum stator d-axis current at 60Nm of torque.

Comparing the reference operation from Fig. 5.5a with the operating points from Fig. 5.5c, the PWM-beat is more prominent due to a lower level of damping by the cage rotor. This causes higher flux and increases the 6th, 8th and 12th saturation harmonics order, but reduces the , ~18th and ~26th mechanical harmonics order, caused by the rotor slotting. The reduction is caused due to the tooth saturation decreasing the electromagnetic slotting effect. The interaction harmonics from the saturation and slotting is partly increased, like the ~24th order, and partly decreased, like the ~30th Order. Where the interaction harmonics increase or decrease depending on the spatial position of the saturation to the slotting harmonics, referring to section 3.2.2 [Bis+17; Bis+16; MVP08].

The simulation of the radial force decomposition can also be diagrammed as the force levels in Fig. 5.6, calculated as the sum of magnitudes over all frequencies for each rotor speed. It is shown, that the radial force level in the minimum d-axis current and subsequently in the minimum flux is always less than the radial force levels of the reference and maximum flux operating, caused by the reduced amplitudes of saturation, PWM harmonics as well as the reduction of the fundamental flux. The calculation of the level of the radial force neglects the damping of the structural dynamics system of the stator and an A-weighting in order to validate the noise perception of the human hearing. However, the force decomposition in the minimum flux operation gives a strong indication for noise reduction.

The structural dynamic of the stator is also important in order to reduce noise emissions and damps the forces depending on the temporal frequency decomposition. Thus, the noise emissions are depending on the ratio of the force excitation and the damping of the structure. Relating to the temporal harmonics of the spatial radial force order, the damping on the system can be defined for the stator structure with a transfer function, as shown in section 4.3. Temporal frequencies in the 0th spatial force mode are mostly damped, but frequencies, which are in the range of 7-11kHz are much less damped than others. This effect is very important to reduce noises. In general, the procedure introduced in this paper relates to reduce force excitation. Reducing high amplitudes of force harmonics, which occur in the range of natural frequencies of the structural dynamic, have the highest potential to reduce the noise emissions of the induction machine [Boe13; Kot+17b; Kot+17a]. Finally, the radial velocity $v_r = v_n$ on the surface of the induction machine is defined with (4.17), referring to section 4.4.

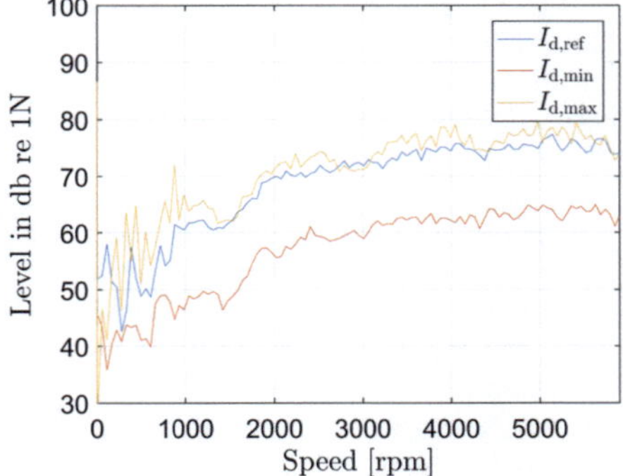

Figure 5.6: Radial force levels for speed run-ups at 60Nm of torque.

5.2 Acoustic Measurements

In order to validate the simulation results from chapter 5.1, measurements are performed. To receive accurate results, the surface velocity of the induction machine is recorded, using an acceleration sensor on the surface of the induction machine. In general, the radial surface velocity v_r shows a direct impact on the sound pressure level with the sound power P [Boe13; Kot+17b], referring to section 4.4, where

$$P(t) \propto |\bar{v}_r(t)|^2, \tag{5.5}$$

with the mean-square surface radial velocity $|\bar{v}_r|^2$ for an infinite axial length defined as,

$$|\bar{v}_r(t)|^2 = \frac{1}{2\pi} \int_0^{2\pi} |v_r(\alpha,t)|^2 \mathrm{d}\alpha. \tag{5.6}$$

The velocity levels in the following, are rated with an A-weighting curve in order to get a ratio of the sound pressure level and the perception of human hearing.

The measured surface velocity levels are shown in Fig. 5.7 for speed run-ups at 20Nm of torque. The reference, minimum and maximum flux operating shows huge differences. As described in chapter 5.1, the minimum flux operation displays the lowest velocity level, but there are also OPs, where the operation in reference flux is below the minimum flux curve. These points can be explained with the Campbell diagrams in Fig. 5.8a - Fig. 5.8c. The saturation for this operating point is lower. Increasing the d-current, increases the flux and the saturation, as well as the PWM-harmonics. This causes a high level of surface velocity for the

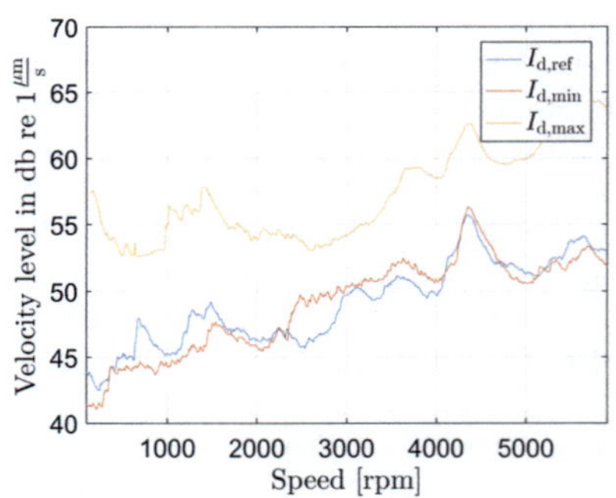

Figure 5.7: Surface velocity levels for speed run-ups at 20Nm of torque.

maximum flux operating. If the minimum flux is operating, the PWM-harmonics, as well as the saturation harmonics are minimized, but the slotting harmonics are increased. If the slotting harmonics operate with the natural frequencies of the system, the slotting harmonics become prominent and increase the velocity level over the reference level.

The measured surface velocity levels for speed run-ups of 60Nm and 100Nm of torque are shown in Fig. 5.9a and Fig. 5.9b. The reference, minimum and maximum flux operating differences are not as big as in the measurements for 20Nm of torque.

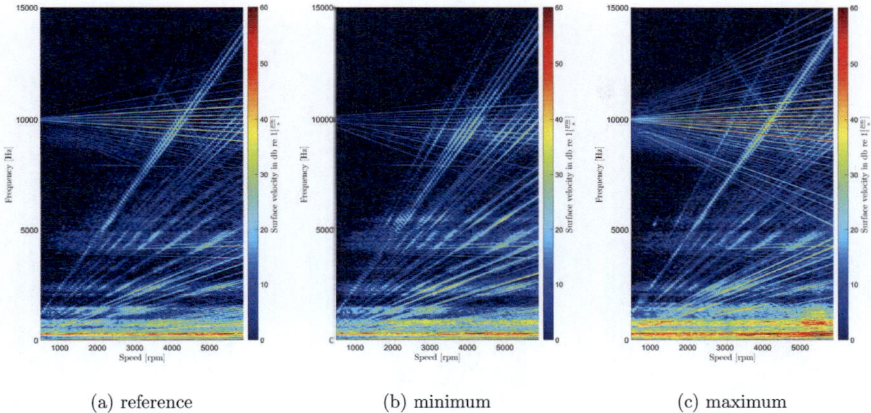

(a) reference   (b) minimum   (c) maximum

Figure 5.8: Surface velocity spectrum for (a) reference,(b) minimum and (c) maximum stator d-axis current at 20Nm of torque.

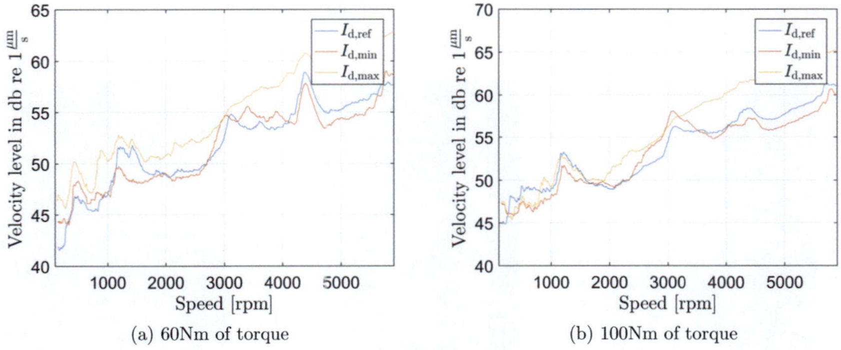

(a) 60Nm of torque   (b) 100Nm of torque

Figure 5.9: Surface velocity levels for speed run-ups at (a) 60Nm and (b) 100Nm of torque.

The theses from chapter 5.1 are also confirmed for these operating points. Relating to the slotting, saturation and PWM-harmonics, the measurements picture the same behavior, as shown for the 20Nm of torque measurements, shown in the Campbell diagrams for 60Nm of torque in Fig. 5.10a - Fig. 5.10c and 100Nm of torque in Fig. 5.11a - Fig. 5.11c. Due to the stator current limitation, the operating range for higher torques gets more and more limited. This in turn reduces the the flux operating points and also the differences between the measurements.

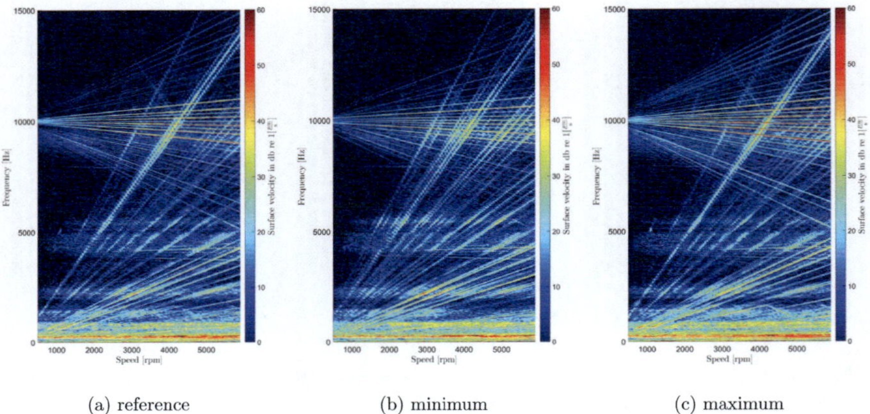

(a) reference   (b) minimum   (c) maximum

Figure 5.10: Surface velocity spectrum for (a) reference, (b) minimum and (c) maximum stator d-axis current at 60Nm of torque.

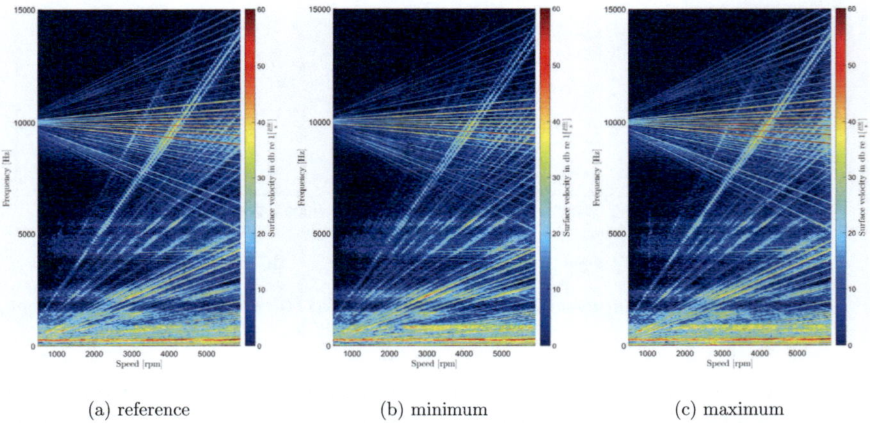

(a) reference   (b) minimum   (c) maximum

Figure 5.11: Surface velocity spectrum for (a) reference, (b) minimum and (c) maximum stator d-axis current at 100Nm of torque.

The potential velocity differences regarding the induction machine surface are shown for 20Nm of torque in Fig. 5.12, for 60Nm of torque in Fig. 5.13a and for 100Nm of torque in Fig. 5.13b. The measurements are performed for different stator d-axis currents, including a bigger range of d-axis operating currents, as shown in the comparison above. This potential analysis shows the maximum reducibility of the surface velocity for each rotor speed. Especially at partial load, the operating area has the biggest influence and subsequently the highest potential for noise reduction. Due to the current limitation, the operating range for a higher torque is limited as well. Nevertheless the operating points offer a chance for reducing the noise emissions for the beginning field weakening area at 3500rpm.

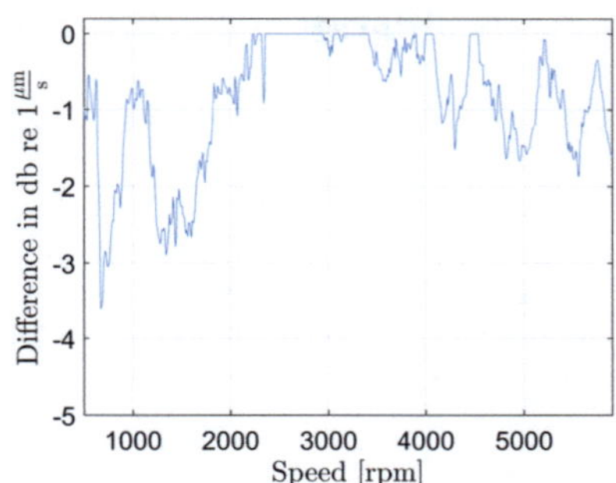

Figure 5.12: Maximum of velocity level difference for different stator d-axis current at 20Nm of torque.

At higher torques, the system temperature increases much faster in comparison to lower torques, influencing the acoustic measurements. This fact blurs the measurement and the potential at higher torques. Shown in Fig. 5.13b, the potential has to be rated lower, than shown especially for speeds above 4000rpm.

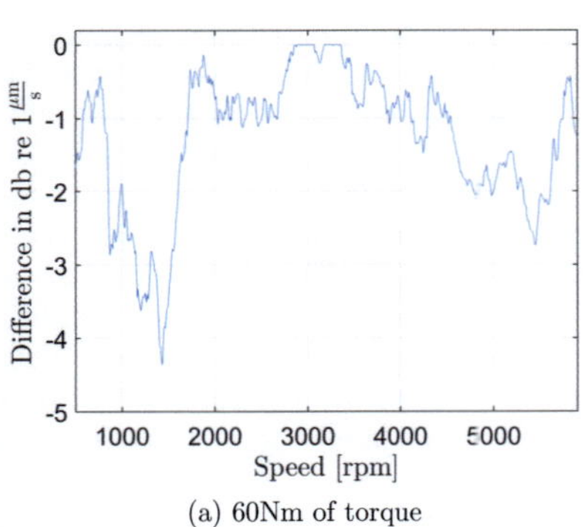

(a) 60Nm of torque

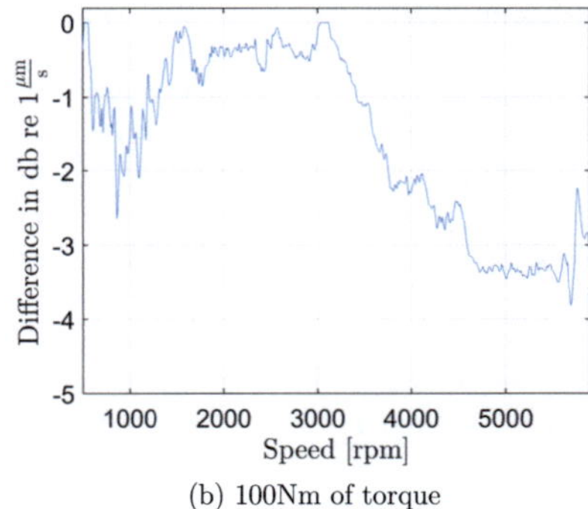

(b) 100Nm of torque

Figure 5.13: Maximum of velocity level difference for different stator d-axis current at (a) 60Nm and (b) 100Nm of torque.

5.3 Calculation of the Operating Points

In order to calculate the complete operating area of the induction machine, this work defined a model. The fundamental machine behavior can be described in the d-q-axis system, which is referencing on the rotor flux [DPV11], referring to section 3.1.1. Additionally, measurements or a vibration syntheses model is applied to optimize the electrical drive. The calculation of the operating points mainly depend on the optimization criteria, that are used for efficiency reasons. This section compares an efficiency-optimized operating strategy to a possible NVH-optimization.

5.3.1 Operating Points for Best Efficiency

In order to calculate the best efficiency of the machine, the machine loss equations from [MVP08; And15; vPSH16], or test bench measurements are applied to calculate the losses for the induction machine operating range. The loss calculation is taken in order to find the minimum of power losses for each torque and stator d-current. The results are shown in Fig. 5.14 for different torques dependent on power losses of the e-drive and the stator d-currents. With the information of stator d-axis currents at different torques, the machine's best efficiency trace is calculated, as shown in Fig. 5.15.

5.3.2 Operating Points for Best NVH-behavior

The calculation for the best NVH-behavior is based on the sound pressure level, or surface velocity measurements, from the sections before. This calculation can also be performed with a vibration syntheses model. It is shown, that the surface velocity and subsequently the sound pressure level can be mainly reduced for low stator d-axis currents. This effect can be explained by the proportionality of the stator d-axis current and the machine magnetization flux [DPV11]. As already explained, in minimum flux regions, the saturation is minimized, whereby the slip frequency and subsequently the rotor current operate at its maximum for the given torque. With the reduction of saturation, saturation harmonics are less stimulated. Increasing the induction into the rotor bars, due to the slip maximum, helps to compensate harmonics in the magnetic flux density in the air-gap of the machine. Noises which occur due to this excitation are hence reduced. Due to the slotting harmonics, the operation within the minimum flux region is not the best solution for every operating point.

For the optimization criteria, the power losses of the electrical drive for different stator d-currents are shown in the minimal flux operating strategy in Fig. 5.14. The operations in the stator d-q-axis system are shown in Fig. 5.15. These diagrams show, that with changing the operating strategy, the efficiency of the e-drive decreases and can cause thermal issues, when permanently operating within the same operating points. This information is very important in order to define the optimization criteria for the NVH-optimization.

As shown in the potential analysis for the maximum surface velocity difference between the three strategies, the optimization criteria are defined as shown in (5.7). In order to reduce the efficiency decrease in case of the flux reduced operating, the strategy for best efficiency operating is not changed till a threshold of the surface velocity difference of -2dB, meaning 20% improvement, is reached. The switching of the different d-axis currents can even be further reduced by adding a second condition. The operating points $I_{d,opt}$ are not changed if there is no operating point nearby, that shows an improvement of -0.5dB, meaning an improvement of 5%, regarding the velocity difference.

$$I_d(t) = \begin{cases} I_{d,\text{eff}}(t) & , \text{ if } \min(v_r(\alpha,t)) > v_r(\alpha,I_{d,\text{opt}}(t)) + 2\text{dB} \\ I_{d,\text{opt}}(t) & , \text{ if } \min(v_r(\alpha,t)) = v_r(\alpha,I_{d,\text{opt}}(t)) + 2\text{dB} \end{cases} \tag{5.7}$$

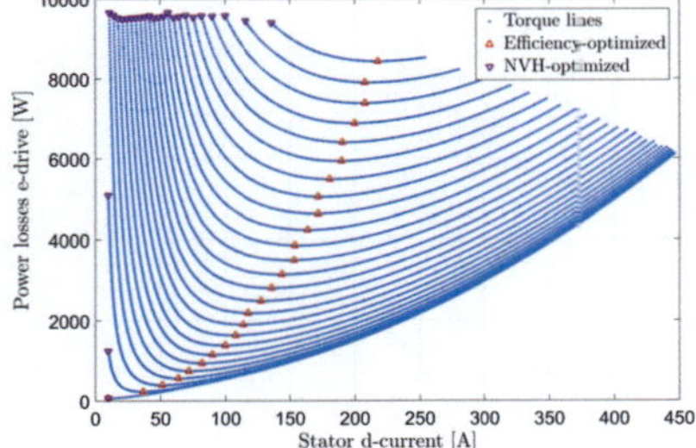

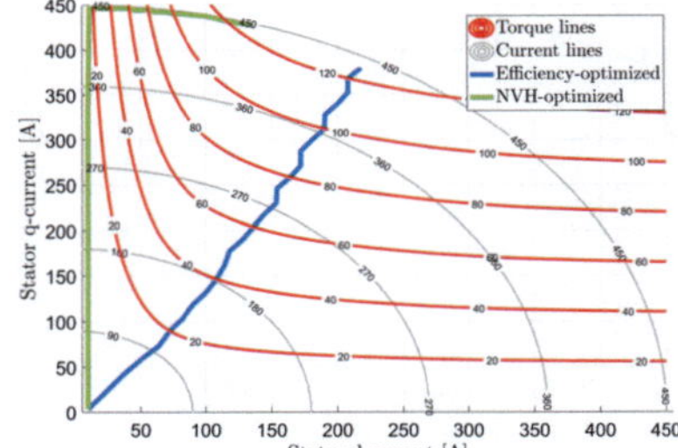

Figure 5.14: Optimized stator d-currents changing with torque reference.

Figure 5.15: Optimized operating points at the fundamental speed range.

The optimized operating is displayed for 20Nm - 60Nm of torque in Fig. 5.16a - Fig. 5.16c and for 80Nm - 120Nm of torque in Fig. 5.16d - Fig. 5.16f. It is shown, how the operating is changed from the NVH-optimized to the efficiency-optimized curves in Fig. 5.15. These operating strategies show an improvement for reaching the field weakening area, because the minimum flux region directly supports this behavior. However, the stator d-axis current is changed very fast, which can have an influence on the dynamics of the system.

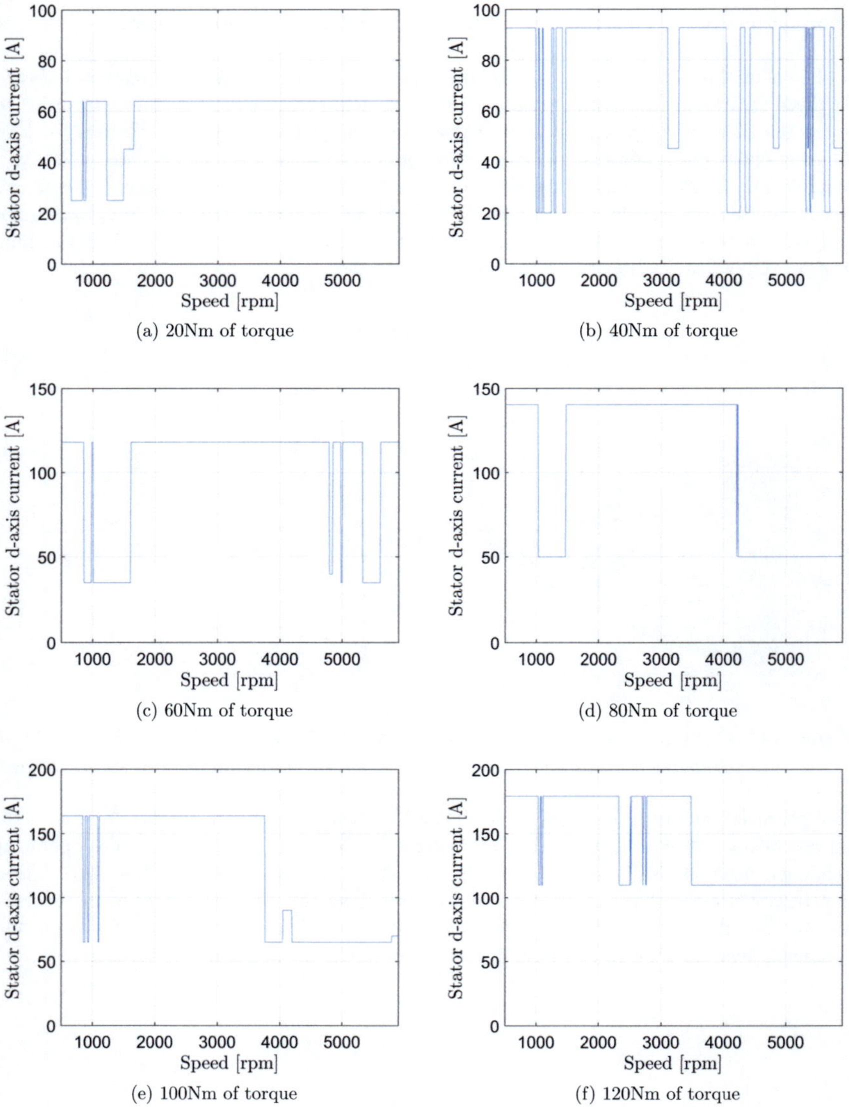

Figure 5.16: Stator d-axis current for noise optimized operating at (a) 20Nm, (b) 40Nm, (c) 60Nm, (d) 80Nm, (e) 100Nm and (f) 120Nm of torque.

5.4 Validation of the Optimization

The optimized strategy is validated with a surface velocity measurement speed run-up at 40Nm of torque. Fig. 5.17 shows the comparison of the measured NVH-optimized and efficiency-optimized strategy. Because of the influences of temperature, the velocity levels at low speeds are not completely matching. After reaching the same operation temperature, the NVH-optimized strategy follows the efficiency-optimized velocity level, but changes the flux level in order to reduce the velocity levels for the given optimization criteria. The operating points 3500rpm and 4300rpm of rotor speed illustrate and support the validity of changing the operating strategy in order to reduce the noise emission of the electrical drive.

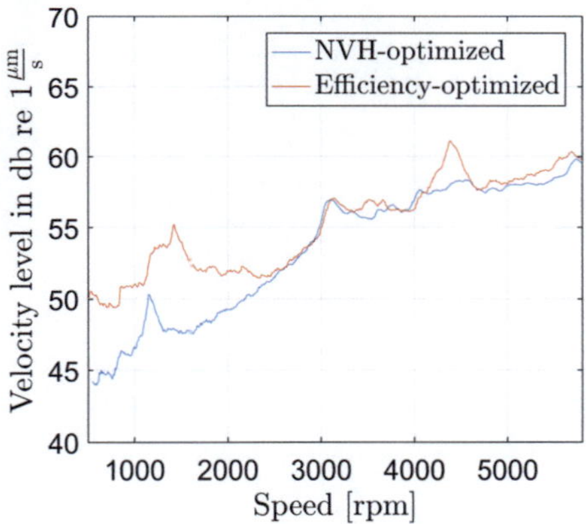

Figure 5.17: Measured surface velocities with NVH- and Efficiency-optimized operating strategies.

5.5 Side Effects of Operating Point Adaption

Changing the operating strategy influences the noise emissions. Additionally, the harmonics content of other quantities like the stator current harmonics, rotor current harmonics as well as the torque ripple are also influenced. Thus, these quantities can also have a huge influence on the NVH behavior of the electrical drive and are examined in more detail in the following sections.

5.5.1 Stator Current

The measured stator current for the reference, minimum and maximum flux operations are diagrammed in Fig. 5.18a - Fig. 5.18c. The harmonics content of the stator current displays the same behavior like the surface velocity, where the saturation and inverter harmonics are minimized in the minimum flux operating point and maximized in the maximum flux operation.

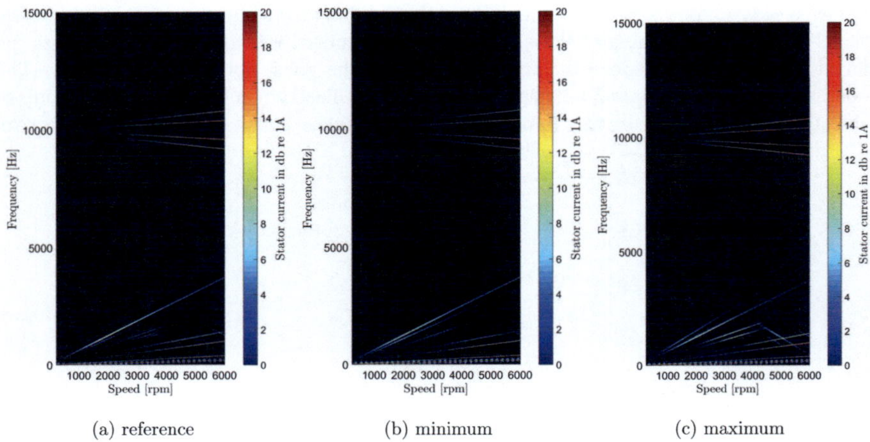

(a) reference   (b) minimum   (c) maximum

Figure 5.18: Measured FT based stator current spectrum at 60Nm of torque for (a) reference, (b) minimum and (c) maximum flux

5.5.2 Rotor Current

The simulated rotor current FT based spectra are diagrammed in Fig. 5.19a - 5.19c for 60Nm of torque. The harmonics content shows the same behavior like the stator current, but the prominence of the slotting harmonics in the minimum flux operation shows a higher visibility.

5.5.3 Electromagnetic Torque

The simulated torque harmonics analysis shows the same expected behavior, in Fig. 5.20a - Fig. 5.20c, as the saturation and inverter harmonics are reduced in the minimum flux operation and increased in the maximum flux operation.

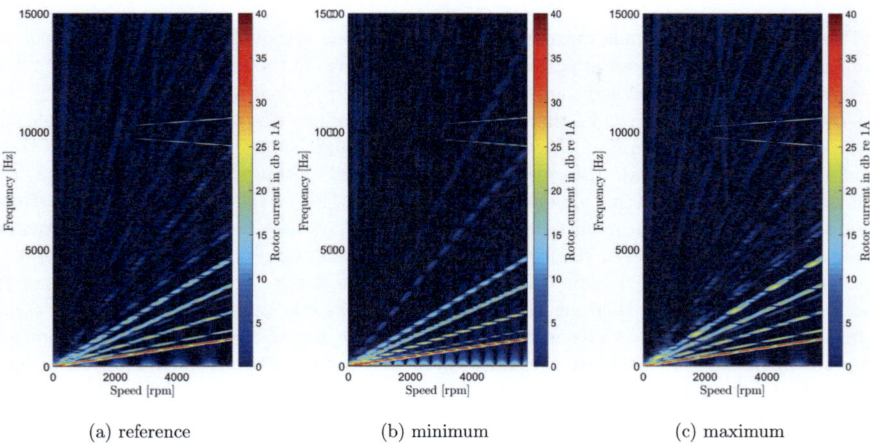

(a) reference (b) minimum (c) maximum

Figure 5.19: Simulated FT based rotor bar current spectrum at 60Nm of torque for (a) reference, (b) minimum and (c) maximum flux

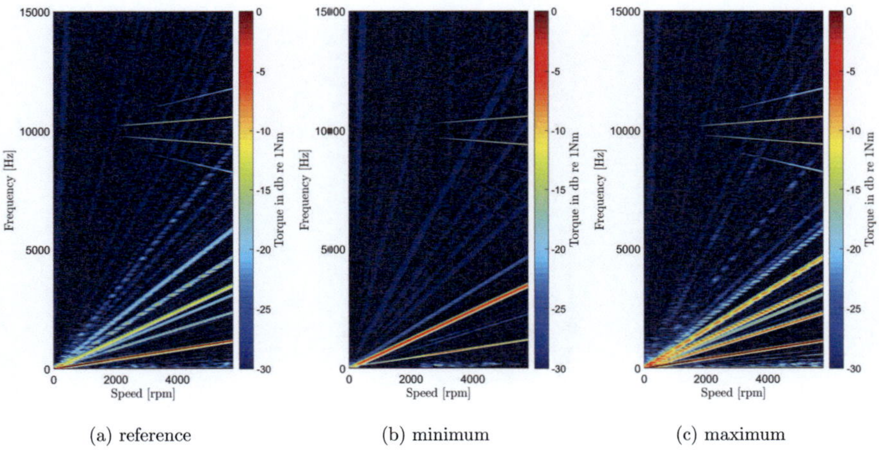

(a) reference (b) minimum (c) maximum

Figure 5.20: Simulated FT based electromagnetic torque spectrum at 60Nm of torque for (a) reference, (b) minimum and (c) maximum flux

5.6 Conclusion

This chapter proposes a new strategy to reduce the noise emissions by changing the induction machine operation without any mechanical design change of the induction machine. As it is shown in chapter 5.1, influencing the electromagnetic force excitation and subsequently the noise emissions, the effects of the occurring harmonics are explainable, and simulable for the operating point adaption. The acoustic measurements support these theories and their validity and help to create the optimization for the best suitable NVH-optimized behavior. It is also shown, that other quantities like the harmonics of the stator and rotor current as well as the torque harmonics are influenced. This can help reducing noise emissions of the coupled transmission or reducing the size of the direct current (DC)-link capacitor. On the other hand, the NVH optimization leads to less efficiency, relating to the optimization criteria and can cause thermal issues. The induction machines' temperature has a big influence on the noise emissions and complicates the calculation of the NVH-optimized operating points. Nevertheless, the cause-effect relationship between the induction machine operating system and the harmonic force stimulation is introduced. Additionally, this work outlines a procedure in order to optimize the electrical drive with regards to the discrepancy of efficiency and noise excitations changing the control strategy.

6 Elimination of Harmonics

After the NVH reduction based on the induction machines' operating point (OP) adaption, in chapter 5, the electrical machines' noise emissions can also be influenced using the method of harmonics injection. For using these methods, the mechanical induction machine design does not have to be changed, because of the extension of the machines' inner cascade control circuit. After the introduction of the principle of the harmonics injection, the different control algorithms are introduced and discussed. It is also necessary to distinguish between the origin of the saturation and slotting harmonics as well as the harmonics caused by the mechanical eccentricities or non-uniformity of the air-gap. For the elimination of the harmonics, the open loop and closed loop control system are shown and compared to each other.

Principle of harmonics elimination

The principle of harmonics is shown in Fig. 6.1, where the characteristics of the electrical machine can be current or torque harmonics as well as surface velocities or noise emissions. With the harmonics injection, a signal is created in order to reduce or eliminate the target harmonics, shown as the resultant characteristics. The harmonics elimination is analogous to the compensation of an occurring signal.

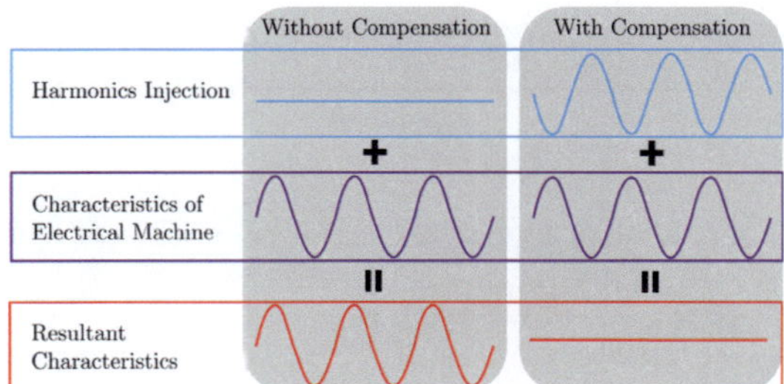

Figure 6.1: Schematic diagram of the harmonics elimination.

Harmonic injection in general

The method of the harmonic injection is generally used to inject harmonic signals into the electrical circuit, e.g. an electrical drive as shown in Fig. 6.2a and Fig. 6.2b, with the fundamental current controller (1), the inverter (2) and the induction machine (3). The stator current is kept at the desired stator current $\vec{i}_\mathrm{s}^{\,*}$. The controller calculates the desired voltage level $\vec{u}_\mathrm{s}$ for the inverter. The harmonics injection is placed at this point as the additional desired harmonics voltage, shown as $\vec{u}_\mathrm{s,h}$. The inverter calculates, based on the given desired voltage, the timing of the IGBTs which sets the three-phase voltage for the induction machine. For the injection method, it is not necessary if the desired harmonic of the electrical drive has to be increased or decreased by the injected signal.

For the harmonics elimination, the suppression of the disturbing orders is supported with the harmonics injection, where the injection does not necessarily have to be decoupled in the control circuit, shown in Fig. 6.2a. For the harmonics creation, the injected harmonics can present themselves as disorders for the control circuit and have to be decoupled, as shown in Fig. 6.2b. The harmonic injection is decoupled either by an induction machine model (4), calculating the additional harmonic current $\vec{i}_\mathrm{s,h}$ and reducing this quantity from the input current of the fundamental PI-controller, or by a low-pass filter (LPF) or a NF (5), as defined in chapter 2.2.2.4, hiding the higher harmonics content of the measured stator current for the fundamental PI-controller. The amplitude and the phase angle of the open loop injected harmonic voltage $\vec{u}_\mathrm{s,h}$ can be estimated or calculated in a pre-calculation. The values are then compiled in a look-up-table (LUT), whereby this procedure can also be performed in closed loop system.

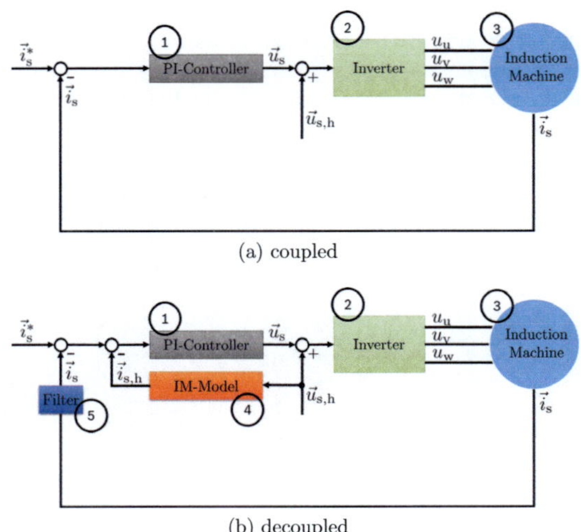

(a) coupled

(b) decoupled

Figure 6.2: Signal flow chart of the open loop (a) coupled and (b) decoupled harmonic signal injection

6.1 Open Loop System: Control of Eccentric Rotor Position

The open loop control relates to quantities, which are hard to be included in the closed loop system. Especially harmonics, which are close to the fundamental frequency can cause instabilities in closed loop systems, as well as harmonics, which are too sophisticated to be included in a series solution without the extension of additional hardware. A special example would be the reduction of harmonics caused by eccentricities, because of their low frequency cause relation, which are entirely performable with the asynchronous stator and rotor fields of the induction machine.

6.1.1 Definition of Eccentric Rotor Positions

In general, there are two different classifications of eccentricities. On the one hand, there is a static eccentricity occurring due to manufacturing imprecisions, which is always present in real electrical machines. This type of eccentricity is defined as a constant eccentric rotor position, which is not completely in the center, or a spatial change of the air-gap due to the ovalization of the stator lamination packet. On the other hand, there is a dynamic eccentricity, which can also

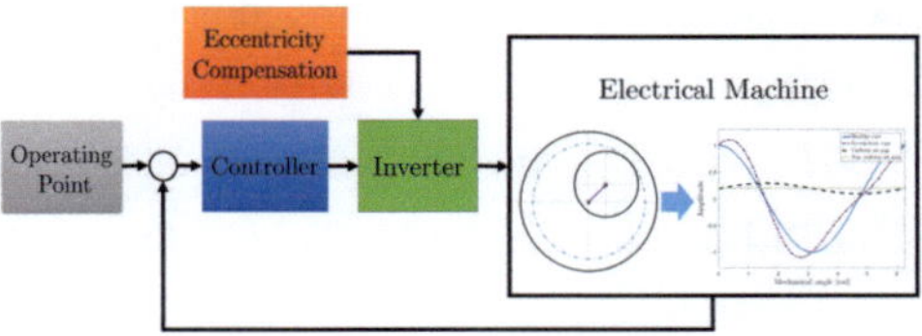

Figure 6.3: Schematic diagram of eccentric rotor position

occur due to manufacturing imprecision, e.g. for an axial inclined position of the rotor or an unbalanced rotor. Static eccentricity can also cause dynamic eccentricity due to the unbalanced magnetic pull [MVP08; HW17; CS16]. At the position where the air-gap is reduced, the rotor is pulled from the rotating magnetic forces and begins to rotate with the magnetic force excitation.

6.1.2 Principle of Rotor Eccentricity Elimination

The occurring harmonics are defined in (6.1) with the stator frequency f_s, number of rotor bars Q_R, slip s and pole pair number of the machine p, and all the occurring harmonics in the magnetic flux density in the air-gap of the induction machine h_B. The variable n_{ecc} is set to 0 for static eccentricity and set to $n_{ecc} \in \mathbb{N}^*$ for dynamic eccentricity [Nan+11; NBT02].

$$f_{ecc} = [(k_n Q_R \pm n_{ecc}) \frac{1-s}{p} \pm h_B] f_s \quad , \text{ with } \quad k_n \in \mathbb{N} \tag{6.1}$$

With these facts, a dynamic eccentricity will rotate with mechanical rotor frequency, convoluted with all the occurring electromagnetic orders and increase the noise excitation. In order to reduce these harmonics, the origin for the noise excitation has to be reduced. In this case, a voltage with the mechanical frequency is injected to act against the fundamental dynamic eccentricity. This order weakens the fundamental magnetic flux density at positions, where the eccentricity is close or directly beneath a pole. Thus, the unbalanced magnetic pull and subsequently the eccentric noise is reduced.

6.2 Closed Loop System: Control on Saturation and Slotting Harmonics

For the closed loop control circuit a harmonics controller is needed in order to calculate the harmonics injection voltage $\vec{u}_{s,h}$ out of the measured stator current $\vec{i}_s$. For this assignment, the following four controllers are compared to each other:

- Repetitive controller in d-q-axis coordinate system (6.2.1)

- Optimization Method with minimization algorithm (6.2.2)

- PI based harmonic controller (6.2.3)

  - PI-controller in the harmonics d-q-axis coordinate system

  - Resonant controller in α-β-axis or d-q-axis coordinate system

6.2.1 Repetitive Controller in d-q-axis Coordinate System

The principle of the repetitive controller is based on the internal model [RCO13a], shown in (6.2), using the ratio $D = T_P/T_s$ of the sample time T_s and the signal cycle duration T_p, in order to create the system response behavior. To control the harmonics of the induction machine, the internal model can be derived to (6.3), with the delay component $W(z)$, the LPF $H(z)$, as well as the harmonic order σ. In order to filter the harmonics content of the current error signal, a HPF has to be defined as well as $G_f(z)$ to guarantee the stability of the system [Che+07; Lim+09; ERD09; RCO13b].

This controller has an excellent steady-state tracking performance and a low total harmonic distortion, based on the high amplification of the fundamental and harmonics [CZQ13; Esc+08]. The normal scope of application is the alternating current (AC) signal systems. The disadvantage of this method is to adapt on varying frequencies. The additional filters and damping factors are also needed in order to guarantee the stability, which impairs the performance of the controller, shown in Fig. 6.5a and 6.5b for a fundamental frequency of 333.3Hz. Other issues are the filter parameters, which are not directly derivable.

$$G_R(z) = \frac{z^{-D}}{1 - z^{-D}} \qquad (6.2)$$

$$IM(z) = \frac{\sigma W(z)H(z)}{1 - \sigma W(z)H(z)} \qquad (6.3)$$

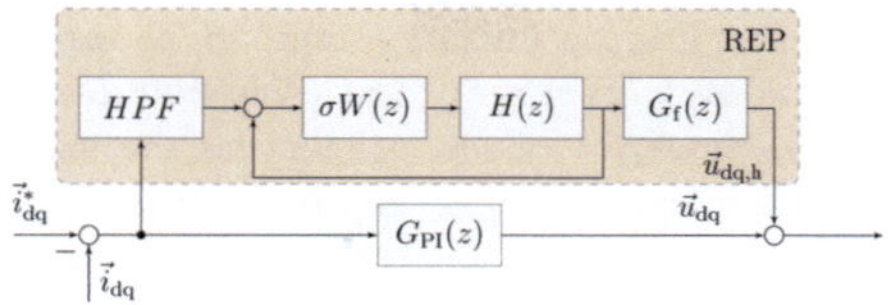

Figure 6.4: Harmonic Repetitive controller in d-q-axis coordinate system

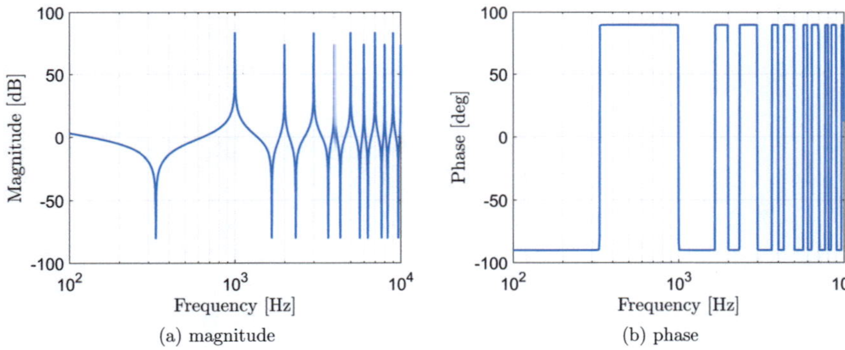

(a) magnitude

(b) phase

Figure 6.5: Transfer behavior of repetitive controller (a) magnitude and (b) phase

6.2.2 Optimization Method with Minimization Algorithm

For using an optimization method, the minimization algorithm is the most important factor for its success. The minimization algorithm tries to minimize the error between the desired and acting value by changing the coefficients of itself and calculates its coefficients for each sampling time [MK96]. In order to control on varying frequencies, adaptive algorithms can be used, which allow to change the bandwidth and the controlled harmonic frequency. For this use case, the adaptive filter [OM03] and adaptive algorithms like a least-mean-square (LMS) algorithm are applied together [MK96; WS85; Sch10]. But these methods are often not used for controlling on currents because of the uncertainty of finding the global minimum of the error field, the often used step size which can cause stability issues and the hard transients behavior because of the need of a window size similar to the repetitive controller.

Equation (6.4a) represents the adaptive filter with weighting factor $\vec{w}_k(n)$ as well as the injection signal $\vec{x}_{dq,h}(n - k)$ and the filter size M. These variables are then implemented into the filter optimization method, shown in Fig. 6.7, using an adaptive algorithm, which minimizes the error $\vec{e}_{\mathrm{dq,h}}(n)$. One example for this minimization algorithm is the so-called LMS, shown in (6.4c), calculating the weighting factor $\vec{w}_k(n+1)$ with the step size μ, injection signal $\vec{x}_{\mathrm{dq,h}}(n)$ and input error $\vec{e}_{\mathrm{dq,h}}(n)$, minimizing the error plane $\xi = \mathrm{E}[e^2]$.

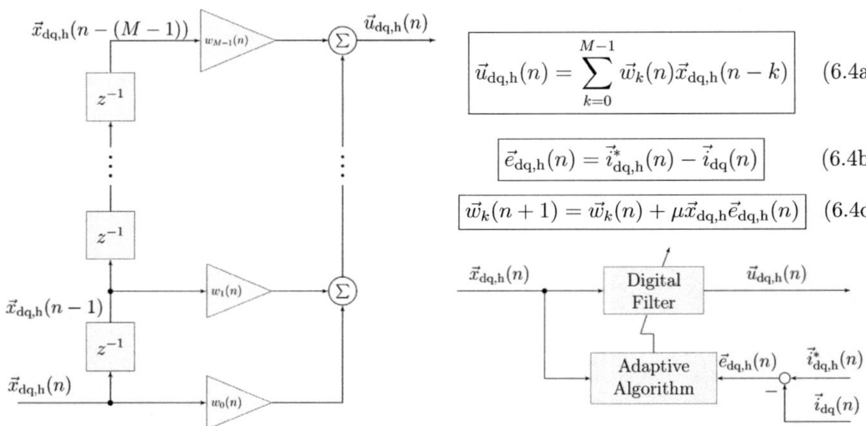

$$\vec{u}_{\mathrm{dq,h}}(n) = \sum_{k=0}^{M-1} \vec{w}_k(n)\vec{x}_{\mathrm{dq,h}}(n-k) \qquad (6.4a)$$

$$\vec{e}_{\mathrm{dq,h}}(n) = \vec{i}^{\,*}_{\mathrm{dq,h}}(n) - \vec{i}_{\mathrm{dq}}(n) \qquad (6.4b)$$

$$\vec{w}_k(n+1) = \vec{w}_k(n) + \mu\vec{x}_{\mathrm{dq,h}}\vec{e}_{\mathrm{dq,h}}(n) \qquad (6.4c)$$

Figure 6.6: Digital adaptive filter

Figure 6.7: Optimization method

6.2.3 Proportional-integral (PI) - based Harmonic Controller

For the PI-based harmonic controller, filter parameters are used like the proportional and integral components of the fundamental PI-controller, shown in (2.27) in chapter 2, and adjusted for its application.

PI-controller in the harmonics d-q-axis coordinate system

The PI-controller is the mostly used type of controller in order to control the electrical machine's fundamental behavior. For the use of this controller as an harmonics controller, the measured stator current is transformed into the d-q-axis coordinate system using the harmonic order h to create the hth multiple of the fundamental flux angle. The advantage of this control strategy is the simple implementation, but depending on the discretization method, it can take a lot of computation time. The error susceptibility increases as well with the application of the rotor flux angle, because of sampling and position error, which is increased by the the high multiple of the measured rotor angle signal [HGP12]. As shown in (6.5), with the parameter $K_{\mathrm{P_h}}^{\mathrm{PI}}$ for the proportional and $K_{\mathrm{I_h}}^{\mathrm{PI}}$ for the integral components, which is the same control method like the mostly used fundamental PI-controller adjusted for the harmonics control, shown in Fig. 6.9.

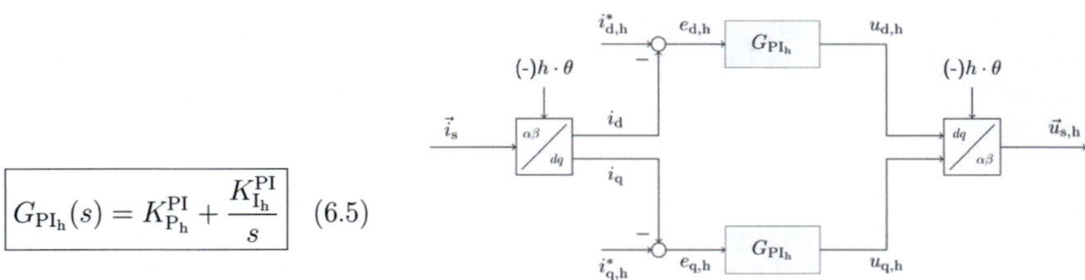

$$G_{\mathrm{PI_h}}(s) = K_{\mathrm{P_h}}^{\mathrm{PI}} + \frac{K_{\mathrm{I_h}}^{\mathrm{PI}}}{s} \qquad (6.5)$$

Figure 6.8: Harmonic PI-controller in d-q-axis coordinate system

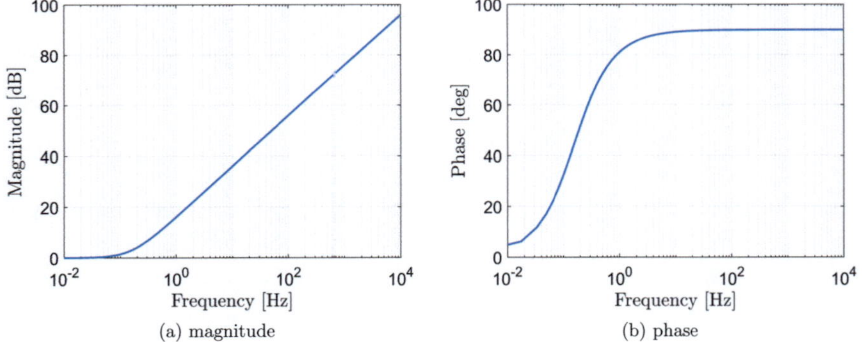

(a) magnitude

(b) phase

Figure 6.9: Transfer behavior of harmonic PI-controller (a) magnitude and (b) phase

Resonant controller in α-β-axis or d-q-axis coordinate system

Resonant controllers are mostly used as active power filters (APFs) in electric power systems. This type of controller consists of two PI-controllers, where one controller is for the positive and the other for the negative encircling component of the signal. The resonant controller is connected in parallel to the fundamental controller and is suitable for sinusoidal signals of any frequencies without steady-state-errors, with the similar transient behavior like the PI-controller. In case of the control of more than one harmonic, the amplification of each controlled harmonic can be set individually. Another advantage of this method is the compensation of delay time, due to the lag time between the input and output signals. The disadvantage of this control strategy is the complexity of discretization, which can cause a higher computation time. In general, resonant controllers are distinguished into proportional-resonant (PR) [Yi+13; ZHB99; Yep+11; HGP12; Yep11] and vector proportional-integral (VPI) [Las+07; Boj+05; Yep11; LTB06] controllers, where both are implementable in the α-β-axis and in the d-q-axis coordinate systems. Using the controller in the d-q-axis coordinate system reduces the complexity of the system by a factor of two [Yep+10; Boj+05; LTB06].

In general, there are two types of resonant controllers: The proportional-resonant (PR) and the vector proportional-integral (VPI) controller, shown as continuous transfer functions in (6.6) and (6.7). Fig. 6.10a and Fig. 6.10b represent the block diagrams of the PR- and VPI-controller, where both need two integrators relating to the proportional-component K_P and integral component K_I as well as the harmonic order h and the fundamental angular frequency $\omega_0 = 2\pi f_0$. The transfer behavior of the two continuous resonant controllers is shown in Fig. 6.11a and Fig. 6.11b, where the amplification of the VPI-controller is much more prominent, but influences the phase of the imminent harmonics content.

$$G_\mathrm{PR}(s) = K_\mathrm{P}^\mathrm{PR} + \frac{K_\mathrm{I}^\mathrm{PR}s}{s^2 + (h\omega_0)^2} \qquad (6.6)$$

$$G_\mathrm{VPI}(s) = \frac{K_\mathrm{P}^\mathrm{VPI}s^2 + K_\mathrm{I}^\mathrm{VPI}s}{s^2 + (h\omega_0)^2} \qquad (6.7)$$

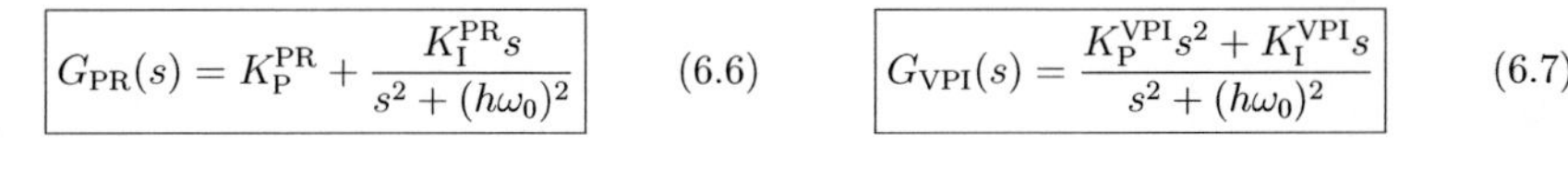

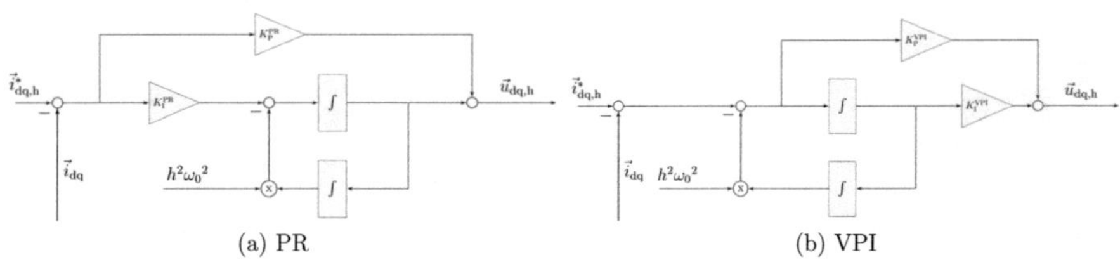

(a) PR (b) VPI

Figure 6.10: Block diagram of (a) PR- and (b) VPI-controller in d-q-axis system

Delay time compensation

Before discretizing the resonant controllers the signal path and the computation time have to be considered in order to create the desired voltages. Thus, the delay time compensation is required, compensating the measured stator current, which represents a departed vector, the computation time of the controllers as well as the timing for the gate drivers and their physical turn-on and turn-off times. For this performance, the resonant controllers get additional terms in order to shift their vectors N times the sampling time T_s. The PR-controller is shown in (6.8), where the proportional-component is not changed due to its

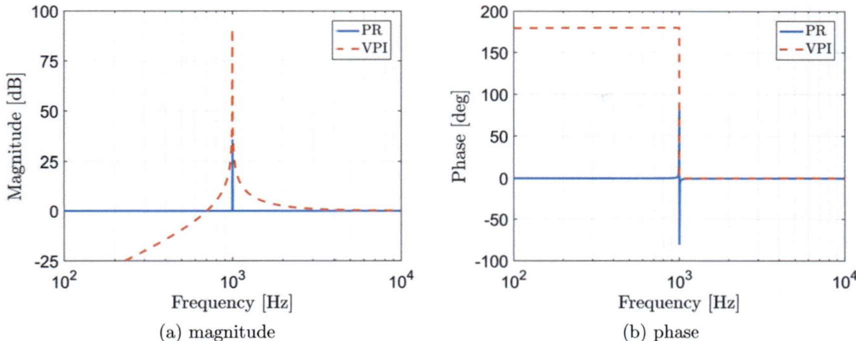

(a) magnitude   (b) phase

Figure 6.11: Transfer behavior of the PR- and VPI-controller (a) magnitude and (b) phase

time independence. The VPI-controller with its delay time compensation is represented in (6.9), where both components are influenced, because of their time dependency.

$$G_{\mathrm{PR}}(s) = K_{\mathrm{P}}^{\mathrm{PR}} + K_{\mathrm{I}}^{\mathrm{PR}}\frac{s\cos(h\omega_0 NT_{\mathrm{s}}) - \omega_0 \sin(h\omega_0 NT_{\mathrm{s}})}{s^2 + (h\omega_0)^2} \tag{6.8}$$

$$G_{\mathrm{VPI}}(s) = (sK_{\mathrm{P}}^{\mathrm{VPI}} - K_{\mathrm{I}}^{\mathrm{VPI}})\frac{s\cos(h\omega_0 NT_{\mathrm{s}}) - \omega_0 \sin(h\omega_0 NT_{\mathrm{s}})}{s^2 + (h\omega_0)^2} \tag{6.9}$$

Discretization of the PI-based controller

After the delay time compensation, the controllers have to be discretized from its continuous function into a discrete function, in order to be applied to the digital systems. The mostly used procedure is the z-Transform, using the z-Transform equivalence. Table 6.1 shows the z-Transform equivalence for the discretization methods applied in this work.

Applying Table 6.1 on the PI-based controllers, the harmonic PI-controller can be discretized using the Euler forward (EF) discretization method as,

$$G_{\mathrm{PI_h}}(s) = K_{\mathrm{P_h}}^{\mathrm{PI}} + \frac{K_{\mathrm{I_h}}^{\mathrm{PI}}T_{\mathrm{s}}}{z - 1}. \tag{6.10}$$

For the resonant controllers with frequency adaption and delay time compensation, three discretization methods are important. The PR-controller is discretized using the impulse invariant IMP method, as shown in (6.11). After the discretization with the delay time compensation, shown in Fig. 6.12a and Fig. 6.12b, with a harmonic frequency of 1kHz and

Table 6.1: Relations of discretization methods (based on [Yep+10])

Method	Equivalence	Notation
First-order hold (FOH)	$F(z) = (1 - z^{-1})Z\left\{\mathcal{L}^{-1}\left\{\frac{F(s)}{s}\right\}\right\}$	$F_{\text{foh}}(z)$
Euler forward (EF)	$s = \frac{z-1}{T_{\text{s}}}$	$F_{\text{ef}}(z)$
Tustin with pre-warping (TP)	$s = \frac{\omega_0}{\tan(\omega_0 T_{\text{s}}/2)}\frac{z-1}{z+1}$	$F_{\text{tp}}(z)$
Impulse invariant (IMP)	$F(z) = Z\left\{\mathcal{L}^{-1}\left\{F(s)\right\}\right\}$	$F_{\text{imp}}(z)$

a sampling rate of 10kHz, the discrete function performs similar to the continuous transfer function. With the delay time compensation, the frequency of the harmonic order can include a very small shift of its frequency.

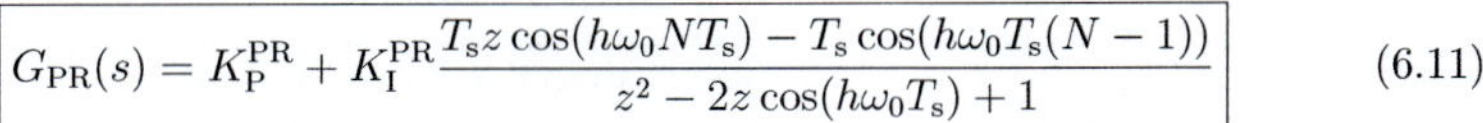

$$G_{\text{PR}}(s) = K_{\text{P}}^{\text{PR}} + K_{\text{I}}^{\text{PR}}\frac{T_{\text{s}}z\cos(h\omega_0 NT_{\text{s}}) - T_{\text{s}}\cos(h\omega_0 T_{\text{s}}(N-1))}{z^2 - 2z\cos(h\omega_0 T_{\text{s}}) + 1} \tag{6.11}$$

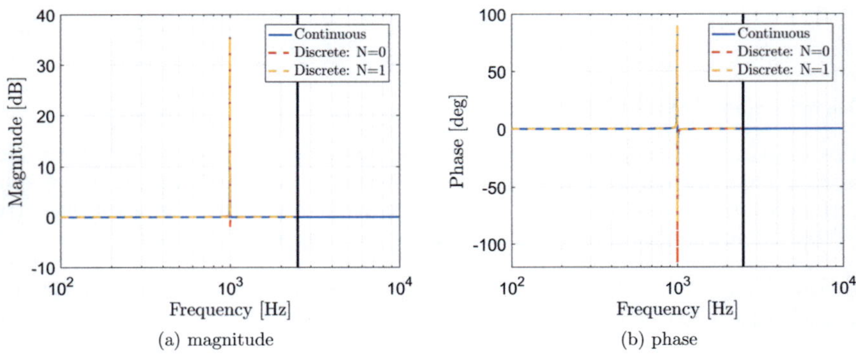

Figure 6.12: Transfer behavior of the PR-controller without and with delay time compensation (a) magnitude and (b) phase

The VPI-controller can be discretized using the first order hold FOH or the tustin with pre-warping TP method for the proportional-component and the impulse invariant IMP method for the integral-component, similar to the PR-controller, represented in (6.12) and (6.13).

$$G_{\mathrm{VPI}}(s) = K_{\mathrm{P}}^{\mathrm{VPI}}\frac{z^2(\sin(h\omega_0 T_{\mathrm{s}}(N+1)) - \sin(h\omega_0 N T_{\mathrm{s}})) - 2z\sin(h\omega_0 T_{\mathrm{s}})\cos(h\omega_0 N T_{\mathrm{s}})}{h\omega_0 T_{\mathrm{s}}(z^2 - 2z\cos(h\omega_0 T_{\mathrm{s}}) + 1)}$$
$$+ K_{\mathrm{P}}^{\mathrm{VPI}}\frac{\sin(h\omega_0 T_{\mathrm{s}}(1-N)) + \sin(h\omega_0 N T_{\mathrm{s}})}{\omega_0 T_{\mathrm{s}}(z^2 - 2z\cos(h\omega_0 T_{\mathrm{s}}) + 1)}$$
$$+ K_{\mathrm{I}}^{\mathrm{VPI}}\frac{T_{\mathrm{s}}z\cos(h\omega_0 N T_{\mathrm{s}}) - T_{\mathrm{s}}\cos(h\omega_0 T_{\mathrm{s}}(N-1))}{z^2 - 2z\cos(h\omega_0 T_{\mathrm{s}}) + 1}$$

$$(6.12)$$

$$G_{\mathrm{VPI}}(s) = K_{\mathrm{P}}^{\mathrm{VPI}}\frac{z^2(0.5(-z^2+1))\sin(h\omega_0 T_{\mathrm{s}})\sin(h\omega_0 N T_{\mathrm{s}})}{z^2 - 2z\cos(h\omega_0 T_s) + 1}$$
$$+ K_{\mathrm{P}}^{\mathrm{VPI}}\frac{(z^2 - 2z + 1)\cos(h\omega_0 T_{\mathrm{s}}/2)^2\cos(h\omega_0 N T_{\mathrm{s}}))}{z^2 - 2z\cos(h\omega_0 T_s) + 1}$$
$$+ K_{\mathrm{I}}^{\mathrm{VPI}}\frac{T_{\mathrm{s}}z\cos(h\omega_0 N T_{\mathrm{s}}) - T_{\mathrm{s}}\cos(h\omega_0 T_{\mathrm{s}}(N-1))}{z^2 - 2z\cos(h\omega_0 T_{\mathrm{s}}) + 1}$$

$$(6.13)$$

The transfer behavior of the discretized controllers without the delay time compensation is similar to the the continuous transfer behavior shown in Fig. 6.13a - Fig. 6.13d, with a harmonic frequency of 1kHz and a sampling rate of 10kHz. Considering a delay time of one PWM period ($T_{\mathrm{s}} = 1/f_{\mathrm{PWM}}$), the transfer behavior changes to a lower drop of harmonics magnitudes prior to the controlled harmonic frequency, changing its phase angle of the signals as well.

6.2.4 Ability of Harmonic Controllers

After the comparison of the different control strategies, the ability of controlling on harmonics have to be assessed. As already mentioned in section 6.2.1, the repetitive controller shows an excellent steady-state performance and a low total harmonic distortion, but challenges for the stable application with varying frequencies. With the fact of a system with varying frequencies, the disadvantages of this method prevail.

The optimization method with the minimization algorithm, introduced in section 6.2.2, with its issues of finding the global minimum of the error field and the step size, which can cause instabilities. Thus, this kind of method is also not useful for the application. Regarding to these facts, the optimization method with the minimization algorithm is also not performed in this application.

In section 6.2.3 the PI-based harmonic controllers are split up into two types of controllers, the PI-controller in the harmonics system and the Resonant controllers. Both types show disadvantages, because of the defining machine parameters dependent proportional- and integral-components, which can relate to the system's temperature, torque and speed. Another disadvantage of the PI-controller in the harmonics system, is the high angle error dependency. Due to the disadvantage of the resonant controllers, the complexity and cause of additional computation time increases, based on their discretization.

-10pt

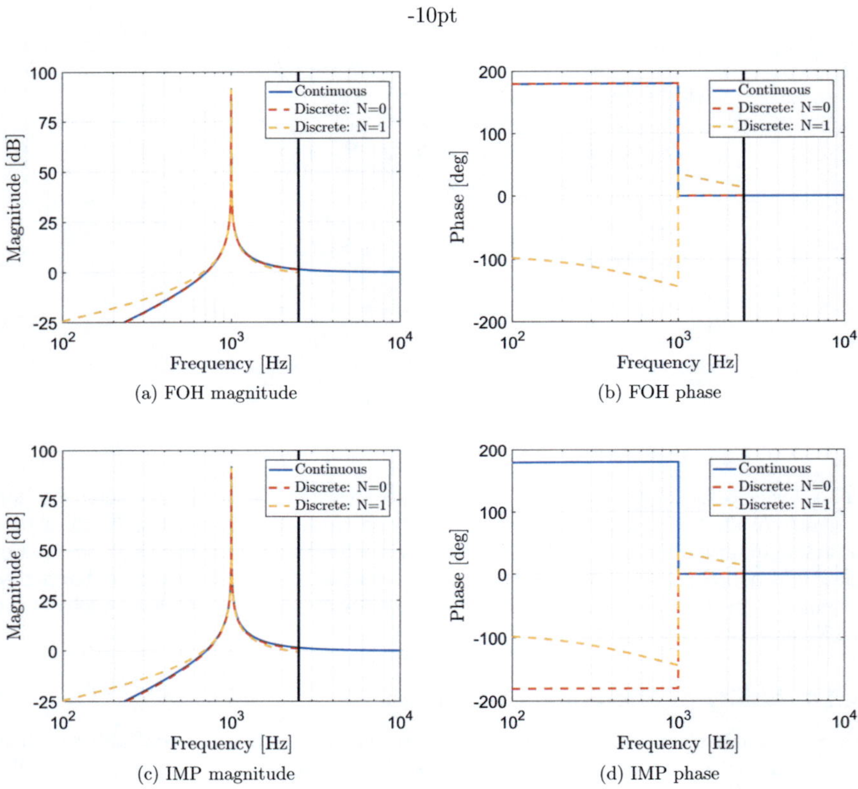

(a) FOH magnitude

(b) FOH phase

(c) IMP magnitude

(d) IMP phase

Figure 6.13: Transfer behavior of the VPI-controller without and with delay time compensation (a) First-order hold (FOH) magnitude, (b) First-order hold (FOH) phase, (c) Impulse invariant (IMP) magnitude and (d) Impulse invariant (IMP) phase.

With the high ability of delay time compensation and adapting on varying frequencies, the resonant controllers are considered and applied in this work.

6.2.5 Principle of Saturation and Slotting Harmonics Elimination

After the definition of the controller, the harmonic order h has to be defined for elimination. With the use of the resonant controllers in the d-q-axis system, the complexity of the system in the α-β-system can be reduced by two. The orders of the saturation harmonics in the

stator current d-q-axis for a three phase induction machine system can be defined with the saturation order ν_{sat} from chapter 3 2.2, as shown in (6.14), with an integer number, because of the synchronicity to the stator frame system. Slotting harmonics are calculated based on the induction machine's slip s, which relating to the rotor frame system, defined in chapter 3.2.2. The slotting harmonics are therefore mainly decimal numbers in the stator frame system, as shown in (6.15). As already shown for the open loop system in Fig. 6.2a and Fig. 6.2b, the fundamental system is controlled using the fundamental current controller (1), the inverter (2) and the induction machine (3).

$$h_{\text{sat,dq}} = 3\nu_{\text{sat}} \qquad (6.14) \qquad h_{\text{slot,dq}} = \frac{Q_r k_\mu}{p}(1 - s) \qquad (6.15)$$

For harmonics injections into the closed loop system, shown in Fig. 6.14a and Fig. 6.14b, the harmonics elimination and subsequently the suppression of the disorders are performed using the harmonics controller (7). Similar to the open loop system, the additional injection does not necessarily have to be decoupled in the control circuit, shown in Fig. 6.14a. For the harmonics creation, the injected harmonics can present themselves as disorders for the control circuit and have to be decoupled, as shown in Fig. 6.14b. The harmonic injection is decoupled either by an induction machine model (4), calculating the additional

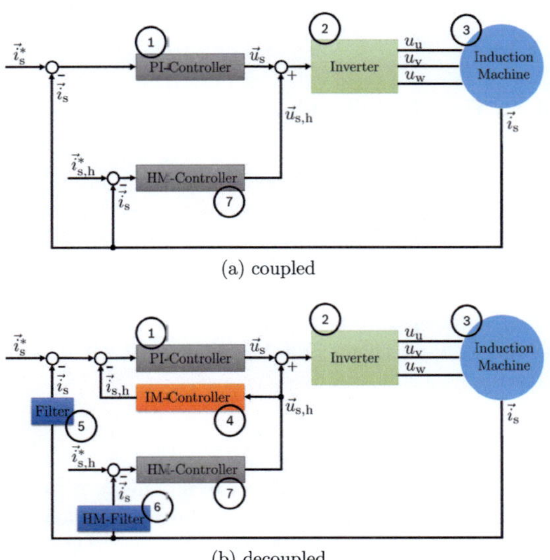

Figure 6.14: Signal flow chart of the closed loop (a) coupled and (b) decoupled harmonic signal injection

harmonic current $\vec{i}_{s,h}$ and reducing this quantity from the input current of the fundamental PI-controller, or by a low-pass filter (LPF) or a NF (5), as defined in chapter 2.2.2.4. This covers the higher harmonics content of the measured stator current for the fundamental PI-controller, similar to the open loop system. In order to decouple the fundamental from the harmonics controller, the harmonic filter (6) can be added by a high-pass filter (HPF) or PF, as also defined in 2.2.2.4, in addition.

6.3 Validation and Comparison

Based on the electromagnetic models from chapter 3 and the structural dynamic model from chapter 4 the control algorithms are tested and implemented in the test bench software. This chapter introduces the test bench results and compare of the proposed methods. Beginning with the control on eccentricities, using the harmonic injection in an open loop system in section 6.3.1, the surface velocity level for an eccentric rotor position of $\sim$10% and $\sim$40% - 50% is compared for different voltage levels and injection angles. After this comparison the closed loop control algorithms are compared in section 6.3.2 for the elimination of saturation and slotting harmonics with the PR-controller and VPI-controller.

6.3.1 Control on Eccentricity Harmonics

Figure 6.15 and 6.16 shows the measured velocity on the surface of the induction machine at 6500 rpm of speed and 20 Nm of torque, depending on the injected voltage and injection direction, for $\sim$10% and $\sim$40% - 50% of eccentricity. The voltage injection begins at the red marked line. The compensation voltage is varied in half volt steps to two volts. The compensation angle is varied from 0 to 360 mechanical degree. The measurement in Fig. 6.15 shows, that there is less influence on the low eccentric machine case. There are variations in the level due to minimal improvements and degradations of the inaccuracy of around 10% and influences of the fundamental controller.

The figure shows, that there are always two global maximum and two global minimum for each voltage level. This is caused by the pole pair number of $p = 2$ of the induction machine and the injection direction of the voltage. They are minimum, if the voltage is injected to reduce the magnetic flux density and to reduce the force excitation. Maximum occur due to the amplification of the magnetic flux density. The measurement with $\sim$40% - 50% of eccentricity, in figure 6.16, shows a higher initial velocity than the measurement with around 10% eccentricity, caused by the eccentric noise. By increasing the compensation voltage and changing the injection direction, it is shown, that the velocity and subsequently the sound pressure level can be reduced.

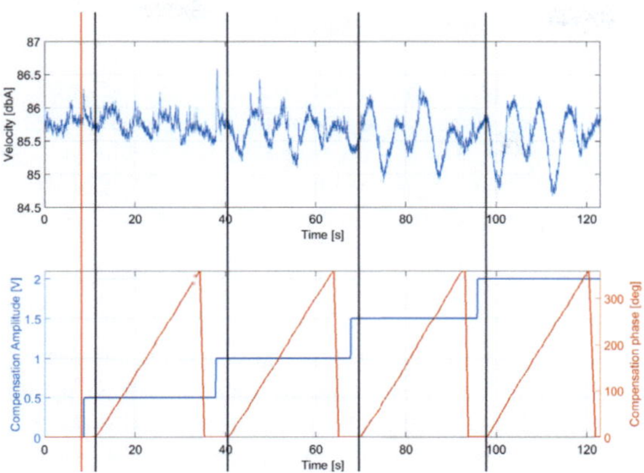

Figure 6.15: Measurement with voltage injection at $\sim 10\%$ of eccentricity

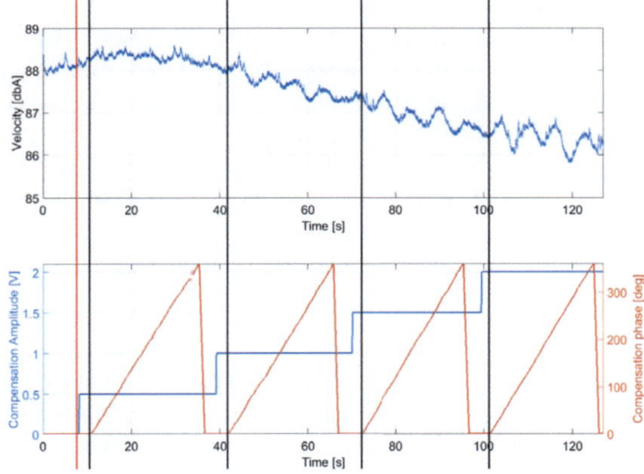

Figure 6.16: Measurement with voltage injection at $\sim 40\%$ - 50% of eccentricity

6.3.2 Control on Saturation and Slotting Harmonics

In this investigation, the PR- and VPI controllers in the closed loop system are compared exclusively for control on saturation harmonics and exclusively for the control on slotting harmonics. In the second step, the comparison is performed controlling on the saturation and slotting harmonics at the same time. For the proportional and integral components of the controllers, the fundamental values are obtained.

6.3.2.1 Compensation of Saturation Harmonics

For the reduction of the saturation harmonics, the harmonics of order $h = 6$ are considered in the stator frame d-q-axis stator current system, in order to control on the 5th and 7th harmonic in the stator current. For this investigation, the battery voltage is set to 250V and 3000rpm of speed, varying the torque level from 0Nm to 40Nm in steady-state operations. Thus, the observed frequency of the 5th, 6th and 7th orders are $\sim$500Hz, $\sim$600Hz and $\sim$700Hz

Stator current

The FT based stator current spectrum is shown in Fig. 6.17a - Fig. 6.17f, for the cases with and without compensation. It is shown, that the 5th and 7th harmonic can be completely compensated with the PR- and VPI-controller. But especially the 11th and 13th as well as the 17th and 19th harmonics increase with both methods, where the VPI-controller causes higher amplitudes for the higher torque region. Thus, the suppression of the 5th and 7th stator current harmonics increase other harmonics, caused by the change of the magnetic flux density caused by the voltage injection. Without the compensation, the magnetic flux density induces the 5th and 7th harmonics into the stator, which causes a field counteracting to its origin, reducing a multiple number of harmonics. With the compensation, the induction of the harmonics is eliminated, and a counteracting field is not built, which causes the higher amplitudes of the higher harmonics. In addition, the controllers can change the angle and amplitude of the side harmonics of the controlled frequency. Especially the VPI-controller shows an influence on harmonics, which have a lower frequency than the controlled order, shown in comparison between PR- and VPI-controller in Fig. 6.11a and 6.11b. This can also have an influence on higher harmonics due to the influence on the magnetic flux density and its induction.

Surface velocity

With the compensation of the 5th and 7th harmonics in the stator frame d-q-axis stator current system, the 6th harmonic is eliminated in the surface velocity level, shown in the FT based velocity spectrum in Fig. 6.18a - Fig. 6.18f. When increasing the 11th, 13th, 17th and 19th harmonics in the stator current, the surface velocity level of the higher harmonics is subsequently increased because of the convolution of the magnetic flux density with itself. Thus, an increase of harmonics in the stator current increases the harmonics content of the mmfs, including the magnetic flux density and subsequently the harmonics in the surface velocity, which implies the radiated noises. Comparing the overall velocity levels with and without the saturation harmonics compensation, the velocity level is not significantly reduced.

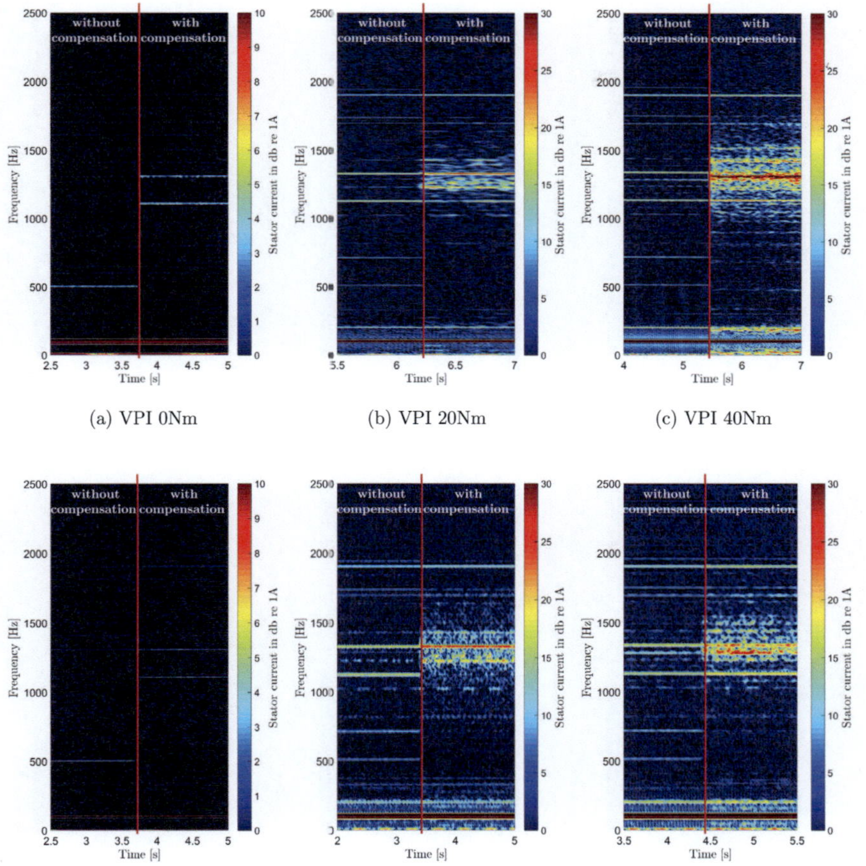

(a) VPI 0Nm (b) VPI 20Nm (c) VPI 40Nm

(d) PR 0Nm (e) PR 20Nm (f) PR 40Nm

Figure 6.17: Measured FT based stator current spectrum of 6th harmonic elimination for (a) VPI 0Nm, (b) VPI 20Nm, (c) VPI 40Nm, (d) PR 0Nm, (e) PR 20Nm and (f) PR 40Nm of torque and 3000rpm of speed

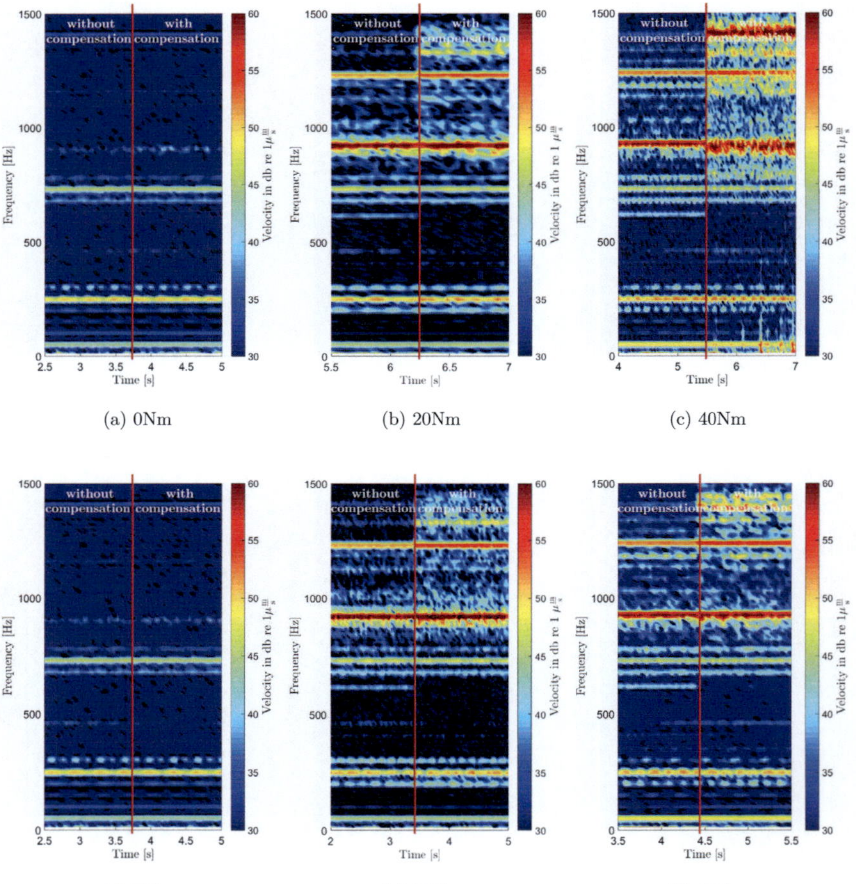

(a) 0Nm

(b) 20Nm

(c) 40Nm

(d) 0Nm

(e) 20Nm

(f) 40Nm

Figure 6.18: Measured FT based surface velocity spectrum of 6th harmonic elimination with VPI controller for (a) 0Nm, (b) 20Nm and (c) 40Nm of torque and 3000rpm of speed

DC-link current

The DC-link current FT based spectrum is shown in Fig. 6.19a and Fig. 6.19b. It is shown, that the compensation of the 5th and 7th harmonics order in the AC currents mostly eliminates the 6th order in the DC-link currents. The comparison of the PR- and VPI-controller doesn't show significant deviations. With this information, the thermal stress of DC-link capacitor is reduced because of the reduced harmonic content in the lower frequency region. Thus, there is a potential to decrease the DC-link capacitance under reserve of the limitations on the DC voltage ripple.

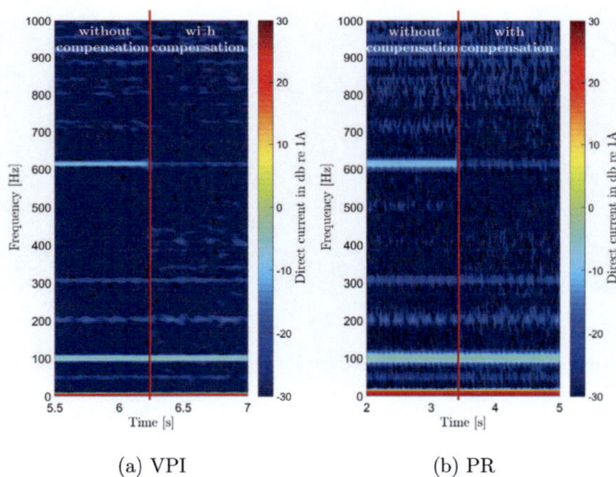

(a) VPI   (b) PR

Figure 6.19: Measured FT based direct current spectrum of 6th harmonic elimination with (a) VPI and (b) PR controller for 20Nm of torque and 3000rpm of speed

6.3.2.2 Compensation of Slotting Harmonics

The compensation of the slotting harmonics is shown for the stator current in Fig. 6.20a. The surface velocity is shown in Fig. 6.20b, the direct current 6.20c for 1500rpm of speed and 20Nm of torque for steady-state operation. For this investigation, the harmonic order is set to 18th order in the rotor frame system, due to the number of rotor slots. With the dependency of the slip frequency and conversion to the d-q-axis stator current stator frame system, the controlled order is defined as $h = 17.1$. Adressing the 16.1th and 18.1th order in the stator current at ~850Hz and ~955Hz. Because of instabilities with the PR-controller, where the PR-controller showed an early instability at around $^1/_5$ of the current control loop sampling rate of 5kHz, this investigation could only be performed with the VPI-controller. When performing the controller at the test bench, a significant dependency on the rotor temperature model was identified, changing the controlled harmonics order in the stator frame system relating to the rotor slip frequency and rotor temperature, estimated by the rotor temperature model.

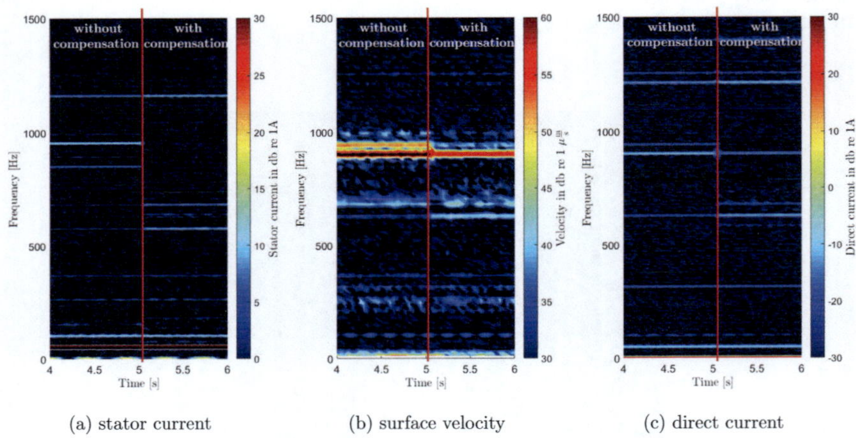

(a) stator current   (b) surface velocity   (c) direct current

Figure 6.20: Measured FT based spectrum of 17.1th harmonic elimination with VPI-controller for (a) stator current, (b) surface velocity, (c) direct current for 20Nm of torque and 1500rpm of speed

Stator current

With the compensation of the 17.1th order in the d-q-axis stator current in the stator frame system, the saturation 5th and 7th stator current order as well as the harmonic combination of slotting and saturation 21.1th order are increased. This also relates to the influence of the VPI-controller on the harmonics on frequencies being lower than the controlled orders as well as the influence of the induced harmonics.

Surface velocity

The compensation of the 16.1th and 18.1th harmonics in the stator current also influences the surface velocity, where the 18th order due to the induced slotting harmonic is completely eliminated and the mechanical slotting harmonic order of 17.1th order is partly compensated. The increased 5th and 7th harmonic in the stator current also result in an increased

6th harmonic in the surface velocity. With this procedure, the surface velocity level is significantly reduced, when compensating on the slotting harmonic orders.

DC-link current

When compensating the harmonics in the stator current, there is also a prominent influence on the DC current, where the DC current behaves analogous to the stator current and surface velocity. The induced 18th slotting order from the magentic flux density is completely eliminated, the mechanical slotting harmonics of 17.1th order is reduced and the saturation 6th order is increased. In the DC current, the 23.1th harmonic relating to the combination of the slotting and saturation harmonics is also increased due to the analogy to the stator current 22.1th order. This procedure doesn't show a significant influence on the overall DC-link current, where the technical design of the DC-link capacitor is barely influenced.

6.3.2.3 Reduction of Saturation and Slotting Harmonics

For the reduction of the saturation and slotting harmonics two harmonic controllers, e.g. two PR-controllers, two VPI-controllers or a combination of a PR- and VPI-controller, are implemented in parallel, in order to apply one controller for the saturation and another for the slotting harmonic orders. When using the d-q-axis implementation procedure, two harmonic controllers are controlling on four harmonic orders in the stator current. In this investigation, the PR- or VPI-controllers at 1000rpm of speed and 20Nm, 40Nm and 60Nm of torque are only applied twice. For the saturation harmonic order, the d-q-axis stator current order in stator frame of 6th order $n = 6$ is used. For the slotting harmonic order, the 18th order in the rotor frame system is applied and changes its frequency in the stator frame system dependent on the electromagnetic torque and subsequently the induction machine slip frequency. The harmonic orders on the d-q-axis stator current in stator frame system are $h = 17.1$, $h = 16.7$ and $h = 16.4$ relating to 20Nm, 40Nm and 60Nm of torque. In order to show the independence of the harmonic controllers in the spectra, the compensation does not exist in the beginning. Then the saturation harmonics are compensated and eventually the saturation and slotting harmonics are compensated.

Stator current

The stator current spectra are shown in Fig. 6.22a - Fig. 6.22c for closed loop control with two VPI-controllers and in Fig. 6.22d - Fig. 6.22f for the closed loop control with two PR-controllers. The compensation with two harmonic controllers in parallel do not show any errors. When turning on the harmonic controller for the saturation order, the combined saturation and slotting order of ~23th order increases, where the amplitude of slotting harmonics slightly decreases. When compensating the slotting harmonic orders in addition, both the controlled saturation and slotting harmonics are eliminated. The VPI-controller shows in this application significant advantages on effecting harmonics higher than

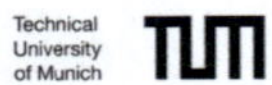

the controlled orders. The caused increased amplitude of the ~23rd order arises due to the saturation harmonic elimination decreases more in comparison to the PR-controllers.

DC-link current

Figure 6.21a and Fig. 6.21b shows the DC-link currents for the compensation of the saturation and slotting harmonics with the VPI and PR controllers at the same time for 20Nm of torque. It is shown, that the 6th harmonics decrease with control on the saturation harmonics, increasing the amplitude of slotting harmonics. After the additional compensation of the slotting harmonics, both the amplitudes of the saturation and slotting harmonics decrease to similar levels like the amplitudes without any compensation. With both controllers, VPI as well as PR, additional harmonics with higher frequencies than the controlled orders are excited.

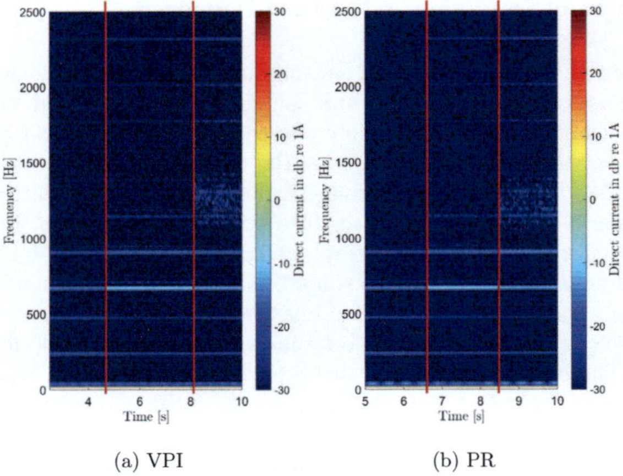

(a) VPI (b) PR

Figure 6.21: Measured FT based stator current spectra of the 6th harmonic in stator frame and 18th harmonic in rotor frame system elimination with (a) VPI and (b) PR controller for 20Nm of torque and 3000rpm of speed

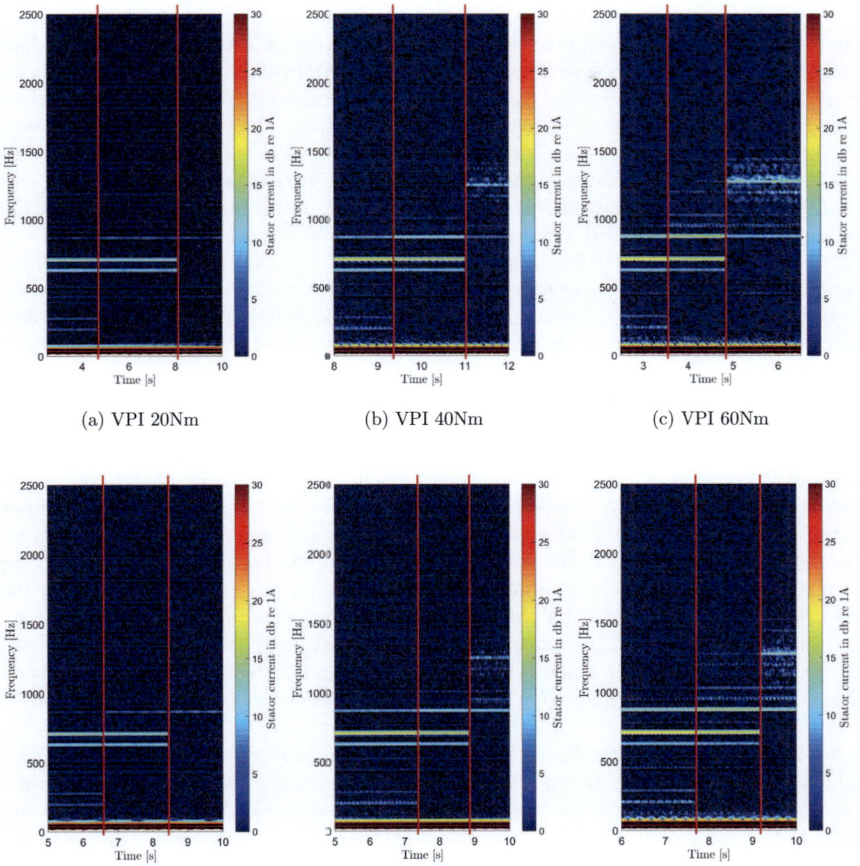

(a) VPI 20Nm (b) VPI 40Nm (c) VPI 60Nm

(d) PR 20Nm (e) PR 40Nm (f) PR 60Nm

Figure 6.22: Measured FT based stator current spectrum of 6th harmonic in stator frame and 18th harmonic in rotor frame system elimination for (a) VPI 20Nm, (b) VPI 40Nm, (c) VPI 60Nm, (d) PR 20Nm, (e) PR 40Nm and (f) PR 60Nm of torque and 1000rpm of speed

6.4 Conclusion

At the beginning of this chapter, the method of harmonic voltage injection is introduced including its implementation as an open loop control system and a closed loop system. The open loop system especially shows advantages for harmonic control frequencies, which are very close to or below the fundamental control frequency. The open loop system can be implemented with LUTs or with a model based observer. The implementation procedures are also introduced in this section, in order to decouple the harmonic injection for the addition or increase of harmonics. For the open loop system, a new control strategy, controlling on eccentricities, is proposed and implemented. The validation section shows, that the unbalanced electromagnetic pull is reduced using different injection voltage levels and angles, where the measured velocity levels for different rates of eccentric rotor positions show a high potential. This procedure, however is rather applied, is mainly useable for the induction machine, caused by its asynchronous rotor frequency compared to the stator frequency.

For the closed loop system, five types of controllers are introduced and compared to each other. As the result of detailed assessments of the harmonic controllers, the resonant controllers are rated as the best suitable solution for this application. These types of controllers are mostly used in electric power systems as APFs and have no steady-state error, an individually adjustable amplification as well as a delay time compensation. Another advantage is the ability to control on the d-q-axis stator current which decreases complexity of the system by two, where one controller controls two stator current harmonics. The disadvantage of this procedure is the complexity of the system equation because of its complex discretization, which is also presented for the used resonant controllers. After the introduction of the harmonics controllers, the influence of the delay time compensation is analyzed, which is needed for the delay time between injection signal and current measurement for higher frequencies and the implementation for decoupling the closed loop system is presented.

After the introduction of the closed loop system and the harmonic controllers, the ability of eliminating saturation and slotting harmonics are evaluated and compared. The elimination of saturation harmonics in the stator current as well as surface velocity and DC-link current with test bench measurements were successfully tested. The simulation results based on the models from chapter 3 and 4 could be confirmed. Controlling the saturation harmonics showed advantages for the DC-link capacitor and the elimination of specific annoying frequencies in particular. However, the overall surface velocity level could not be significantly reduced.

The elimination of the slotting harmonics was successfully tested and confirmed the models from chapter 3 and 4. The slotting controlled harmonic orders were completely eliminated, as the harmonics in the surface velocity and DC-link current were only partly eliminated but mostly reduced. This procedure shows a significant influence on the surface velocity, but with no significant influence noticed on the overall DC-link current.

After the elimination of the saturation and slotting harmonics exclusively, the implementation of two harmonic controllers in parallel was investigated. This procedure was also implemented, successfully tested and shows a high potential to combine the advantages of the saturation and slotting control.

In the end, both the VPI- and PR-controller show high potential in order to reduce harmonics in electrical drives. An improvement of the efficiency of the electrical drive could neither be approved nor disapproved at the test bench measurement. Assessing the single steady-state performance, the PR-controller showed slightly better results, where the VPI-controller showed slightly better results in the steady-state control on two harmonics at the same time. Rating the overall performance especially for the dynamic case, using the delay time compensation, shown in Fig. 6.23a - 6.23c, the VPI-controller shows a better dynamic behavior, where the PR-controller was unstable at ~3000rpm.

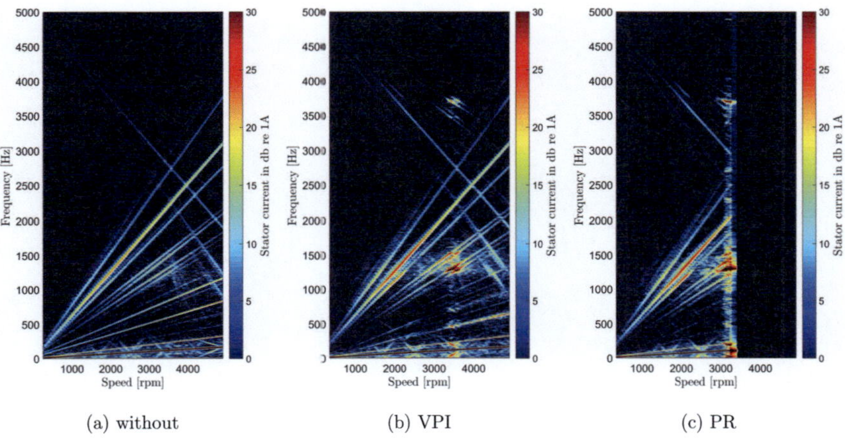

(a) without   (b) VPI   (c) PR

Figure 6.23: Measured FT based stator current spectra of speed run-ups (a) without, (b) VPI and (c) PR 6th harmonic elimination for 60Nm of torque from 250-5000rpm of speed

7 Compendium and Outlook

After the introduction of the construction of Electrical Vehicles (EVs) and explanation of the causality chain of the acoustic noises, as well as the approach for efficient induction machine NVH simulation, in chapter 1, the fundamentals for this work are introduced in chapter 2. Beginning with noise emissions and sources in the electrical drive, the acoustic behavior of electric drive systems is put into context with conventional drive systems and the signal path is identified in general. With the definition of terms in machine acoustics, the signal constraint and characteristics are clarified for the analyzes. Next, this work proposes the Fourier transform (FT) in the spatial frequency domain (SFD) and the temporal frequency domain (TFD), including their display specific formats. Finally the fundamentals of the machine control system and the inverter switching strategy are introduced.

With the proposal of the electromagnetic modeling, in chapter 3, the electromagnetic modeling is split up into the sequential and the integrated electromagnetic simulation. For the sequential electromagnetic simulation the spatial magnetic flux density is calculated using the calculation results of the stator and rotor currents from a fundamental or extended fundamental model. This approach is described in the spatial time domain (STD) and proposed as a new method in the spatial frequency domain (SFD). With this approach, the information from the calculated spatial magnetic flux density is not fed back into the model. With the integrated electromagnetic simulation, a refined model is proposed, where the information in the magnetic flux density is fed back in order to calculate the flux linkages and the induced voltages with the harmonic content of the magnetic flux density. Within that context, the modular air-gap approach similar to an ideal rotating transformer (IRTF) model is proposed in different depth of detail. As it is shown in the comparison and validation section, the approaches are valid and comparable results to FEM simulations and measurements are obtained.

With the calculation of the radial magnetic flux density in the air-gap of the induction machine, the electromagnetic forces have to be put on the structural dynamic model, shown in chapter 4. At the beginning of this chapter, it is shown how the radial forces are spatially and temporally decomposed in order to prepare them for the projection on the stator tooth, where three methods are proposed. With this information, the structural response analysis can be done, based on modal superposition and leading to the force response calculation. The transfer functions (TFs) are created, combining the structural modes with the force modes. For this approach, a new extension is proposed in order to combine the effect for the tooth force projection with the radial forces in a deeper and more detailed way. With these transfer functions (TFs), the surface displacement is processed calculating the surface velocity or surface acceleration. Finally, additions are proposed for the extension of the sound power calculation. The comparison and validation section shows the structural mode shapes are shown as well as the validation of the surface velocity base on test bench measurements. The results emphasize the validity of the model with a good match, even though the results could be potentially improved.

With the electromagnetic and structural dynamic models, investigations for new control strategies are identified and proposed. Chapter 5 proposes a new method, the operating point adaption for induction machines. The idea of this approach is to change the d-q-axis operating point for same machine torque ripple, in order to change electromagnetic behavior and subsequently the forces in the air-gap and with it the machines surface velocity. This approach shows high potential, especially for the reduction of d-axis stator current, which reduces the induction machines flux. The validation of this procedure, based on test bench measurements, showed the potential reducing the machines surface velocity. Additionally, this procedure can also effect the stator and rotor currents, as well the electromagnetic torque positively. With this approach additional challenges have to be regarded, as this procedure could cause thermal issues due to leaving the machines best efficiency operation.

The second approach for the noise optimization is the harmonic voltage injection, proposed in chapter 6. After the open loop voltage injection, where the harmonic excitation of eccentricities caused by eccentric rotor position was successfully reduced, which subsequently reduced the surface velocity, controllers for a closed loop application were proposed. For a closed loop application, the ability of controllers, controlling on saturation and slotting harmonics are compared and an implementation with resonant controllers is developed. The comparison showed, that the proportional-resonant (PR)- and vector proportional-integral (VPI) resonant controllers performed better with regards to their steady-state operation, the individually adjustable amplification, the useability of controlling on the d-q-axis stator current, in order to reduce the number of controllers, as well as the ability of delay time compensation caused by the delay time of measured and injected signals. In order to validate the simulation results, the resonant controllers were implemented and tested at the test bench. In these investigations, the saturation harmonics and slotting harmonics were exclusively successfully eliminated, where finally both harmonics were successfully eliminated in the stator current at the same time. When eliminating these harmonics, additional harmonics were excited, as the potential of reducing noises was the highest for controlling on slotting harmonics. Additionally the elimination of saturation harmonics showed positive effects on the DC-link current. Comparing the PR- and VPI-controller, the PR-controller showed slightly better results at the steady-state operation, because of the reduced excitation of additional harmonics. In this comparison, the VPI-controller showed better results in the transient behavior as well as the ability of controlling on frequencies closer to the sampling rate of the current loop. An improvement of the efficiency, due to the reduction of harmonics, could not be observed.

With the electromagnetic and structural dynamic model, the causality chain of the harmonics and NVH behavior of electrical drives can be deeper understood, new control strategies for better NVH behavior can be developed and design rules for low noise design can be derived. Based on this work, the integrated electromagnetic simulation can be transformed into the spatial frequency domain (SFD), with the proposed procedure from the sequential electromagnetic simulation, reducing the simulation time and increasing the accuracy. The proposed procedures for the NVH optimization could be refined as well. Especially for the operating point adaptation, observers have to be developed for more detailed and accurate results, where the resonant controllers have to be trimmed for controlling on higher frequencies.

In the future, the calculation could be included by the implementation of the electrical drive in a complete electrical axle, with the transmission and its additional harmonics. This potentially helps to build an even higher system knowledge, improving the drive system and increasing the added value.

Appendix

A Induction Machine Drive Data

- Maximum torque: 133Nm

- Maximum speed: 18000rpm

- Maximum power: 70kW

- Continuous power: 35kW

- DC-Link voltage: 180V-250V

- Maximum stator current: 500A

- Pole pair number p: 2

- Number of phases m: 3

- Number of stator slots Q_s: 48

- Number of rotor slots Q_r: 36

- Axial length l_Fe: 112mm

- Outer stator diameter $d_\text{s,out}$: 166mm

- One-side air-gap δ: 0.6mm

- Stator winding: Distributed, zoned, unchorded

- Stator winding material: Copper

- Rotor winding: Cage-rotor, unskewed

- Rotor bar material: Aluminum

B Signal Flow Chart

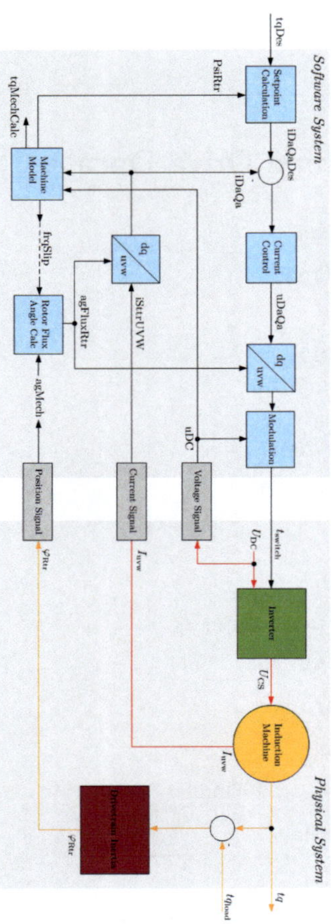

Figure B.1: Signal flow chart of the controlled system

C Comparison of the Electromagnetic Simulation Methods

Table C.1: Electromagnetic simulation methods (based on [ENG13])

	Permeance model	FEM model
Simulation time	++	- -
Accuracy	0*	++
Tangential field calculation	-	+
Current harmonics	0*	+
Saturation calculation	0*	++
Skewing, eccentricities, air-gap abnormalities	0*	++
Fault reactions	+	+
Inverter, controller & sensor influence	++	-
Modeling effort	+	0
Causal relations	++	0

* rating can be improved with the integrated electromagnetic simulation model

Scale: - - very hard, - hard, 0 acceptable, + good, ++ very good

D Park Transformation

$$\begin{bmatrix} u_\mathrm{d} \\ u_\mathrm{q} \end{bmatrix} = \sqrt{\frac{2}{3}} \underbrace{\begin{bmatrix} cos(\theta_\mathrm{R}) & cos(\theta_\mathrm{R} - \frac{2\pi}{3}) & cos(\theta_\mathrm{R} + \frac{2\pi}{3}) \\ -sin(\theta_\mathrm{R}) & -sin(\theta_\mathrm{R} - \frac{2\pi}{3}) & -sin(\theta_\mathrm{R} + \frac{2\pi}{3}) \end{bmatrix}}_{T_\mathrm{P}} \begin{bmatrix} u_\mathrm{U} \\ u_\mathrm{V} \\ u_\mathrm{W} \end{bmatrix} \tag{D.1}$$

$$\begin{bmatrix} i_\mathrm{U} \\ i_\mathrm{V} \\ i_\mathrm{W} \end{bmatrix} = \sqrt{\frac{2}{3}} T_\mathrm{P}^{-1}(\theta_\mathrm{R}) \begin{bmatrix} i_\mathrm{d} \\ i_\mathrm{q} \end{bmatrix} \tag{D.2}$$

E Clarke Transformation

$$\underbrace{\begin{bmatrix} u_\alpha \\ u_\beta \end{bmatrix}}_{\vec{u}_\mathrm{s}} = \frac{2}{3} \underbrace{\begin{bmatrix} 1 & -\frac{1}{2} & -\frac{1}{2} \\ 0 & \frac{\sqrt{3}}{2} & -\frac{\sqrt{3}}{2} \end{bmatrix}}_{T_\mathrm{C}} \underbrace{\begin{bmatrix} u_\mathrm{U} \\ u_\mathrm{V} \\ u_\mathrm{W} \end{bmatrix}}_{\vec{i}_\mathrm{s}} \tag{E.1}$$

$$\begin{bmatrix} i_\mathrm{U} \\ i_\mathrm{V} \\ i_\mathrm{W} \end{bmatrix} = T_\mathrm{C}^{-1} \begin{bmatrix} i_\alpha \\ i_\beta \end{bmatrix} \tag{E.2}$$

F Derivation of Occurring Spatial Orders

Table F.1: Spatial convolution of the air-gap permeance and saturation permeance orders

$\Lambda(u,t)$		Orders of $\Lambda_\mathrm{airgap}(u,t)$								
		-24	-18	-12	-6	0	6	12	18	24
Orders of $\Lambda_\mathrm{Sat}(u,t)$	-4	-28	-22	-16	-10	-4	2	8	14	20
	-2	-26	-20	-14	-8	-2	4	10	16	22
	0	-24	-18	-12	-6	0	6	12	18	24
	2	-22	-16	-10	-4	2	8	14	20	26
	4	-20	-14	-8	-2	4	10	16	22	28

Table F.2: Spatial convolution of the resulting mmf and the saturated permeance orders

$B_r(u,t)$		Orders of $\Theta(u,t)$								
		-23	-17	-11	-5	1	7	13	19	25
Orders of $\Lambda(u,t)$	-10	-33	-27	-21	-15	-9	-3	3	9	15
	-8	-31	-25	-19	-13	-7	-1	5	11	17
	-6	-29	-23	-17	-11	-5	1	7	13	19
	-4	-27	-21	-15	-9	-3	3	9	15	21
	-2	-25	-19	-13	-7	-1	5	11	17	23
	0	-23	-17	-11	-5	1	7	13	19	25
	2	-21	-15	-9	-3	3	10	15	21	27
	4	-19	-13	-7	-1	5	11	17	23	29
	6	-17	-11	-5	1	7	13	19	25	31
	8	-15	-9	-3	3	9	15	21	27	33
	10	-13	-7	-1	5	11	17	23	29	35

Table F.3: Spatial convolution of the magnetic flux density orders with itself

$F_r(u,t)$		Orders of $B_r(u,t)$								
		-9	-7	-5	-3	1	3	5	7	9
Orders of $B_r(u,t)$	-9	-18	-16	-14	-12	-8	-6	-4	-2	0
	-7	-16	-14	-12	-10	-6	-4	-2	0	2
	-5	-14	-12	-10	-8	-4	-2	0	2	4
	-3	-12	-10	-8	-6	-2	0	2	4	6
	1	-8	-6	-4	-2	2	4	6	8	10
	3	-6	-4	-2	0	4	6	8	10	12
	5	-4	-2	0	2	6	8	10	12	14
	7	-2	-0	2	4	8	10	12	14	16
	9	0	2	4	6	10	12	14	16	18

G Comparison of the Electromagnetic Simulation Results

Table G.1: Overview of the compared simulation results

	Fundamental		THD [%]	
Operating point (OP)	1	2	1	2
Speed [rpm]	1000	6000	1000	6000
Fundamental model: Torque [Nm]	132.7	48	0	0
Fundamental model: Stator current [A]	641	290.1	0	0
Fundamental model: Rotor current [A]	1103	467.2	0	0
LIDAG without saturation: Torque [Nm]	131.6	46.9	0.6	0.79
LIDAG without saturation: Stator current [A]	644.2	304.1	0.93	1.8
LIDAG without saturation: Rotor current [A]	1056	467.2	19.9	19.48
LIDAG with saturation: Torque [Nm]	131.6	46.9	2.54	3.4
LIDAG with saturation: Stator current [A]	653.2	297.2	1.92	4.38
LIDAG with saturation: Rotor current [A]	1061	449	21.72	20.95
FEM model: Torque [Nm]	133.5	46	3.77	5.85
FEM model: Stator current [A]	666.3	292.8	1.64	2.53
FEM model: Rotor current [A]	1025	435.6	17.95	21.46

H Constant Tooth Force Approach (tooth)

The constant tooth force approach (tooth) is directly calculable in SFD. The basis equation is shown in (H.1), changing the radial force $F_\mathrm{r}(u)$ with the pseudo inverse transformation matrix $H^{+}_\mathrm{tooth}(u)$ to the tooth force $F_\mathrm{tooth}(u)$ for each time step [Boe13; Gar+97].

$$F_\mathrm{tooth}(u) = H^{+}_\mathrm{tooth}(u)F_\mathrm{r}(u) \tag{H.1}$$

The transformation matrix H_tooth is defined with the radiant ratio of tooth to tooth and slot $c_\mathrm{ratio} = \tau_\mathrm{tooth}Q_\mathrm{s}/2\pi$, using stator slot number Q_s, as

$$H_\mathrm{tooth}(u) = \begin{cases} c_\mathrm{ratio}\,\mathrm{sinc}\left(\frac{\tau_\mathrm{tooth}}{2\pi}u\right) & \text{, where } u = u + k_\mathrm{t}Q_\mathrm{s},\ k_\mathrm{t} \in \mathbb{Z} \\ 0 & \text{, otherwise} \end{cases}. \tag{H.2}$$

I Modal Analysis

This section introduces the structural dynamic eigenvalue problem, involving a derivation of eigenfrequencies, mode shapes as well as the eigenvectors, which are called structural modes, or structural shapes for the electrical machines' structural model. The mode shapes and eigenfrequencies are derived by solving an eigenvalue problem associated to the equations of motion for an undamped, load-free structure, given in a Finite-Element-formulation by a set of N-dimensional coupled, undamped mass-spring differential equations for free vibrations, as shown in (I.1),

$$M\ddot{u}(t) + Ku(t) = 0, \tag{I.1}$$

where u is the vector of structural displacements for the structural degrees of freedom, M is the mass matrix and K is the stiffness matrix. Using a time-harmonic approach $u(\alpha,t) = \phi_{\mathrm{m}}(\alpha)\cos(\omega_{0,\mathrm{m}}t + \varphi)$, where $\omega_{0,\mathrm{m}}$ is the N-th eigenfrequency and $\phi_{\mathrm{m}}(\alpha)$ is the mode shape of the m-th vibration of the system, deriving the algebraic matrix equation

$$-\omega^2 Mu(\omega) + Ku(\omega) = 0 \tag{I.2}$$

with the non-trivial solutions $u(\omega)$ if,

$$\det(K - \lambda M) = 0, \text{ where } \lambda = \omega_0^2. \tag{I.3}$$

The eigenvalue problem from (I.3) can be solved using a structural FEM for a reduced set of $M < N$ eigenfrequencies $\omega_{0,\mathrm{m}}$. Inserting the solution of (I.3) into the expression (I.2), delivers a set of orthogonal, generalized modal displacement vectors, which approximate in sum the real physical behavior of the mechanical structure.

J Harmonic Analysis

After the modal analysis (I), neglecting the damping of the structure, the damping matrix C is considered in this section in order to achieve a mathematical expression for equalizing an external force $F(t)$ shown in TTD in (J.1a) and TFD in (J.2).

$$M\ddot{u}(t) + C\dot{u}(t) + Ku(t) = F(t) \tag{J.1a}$$
$$-\omega^2 Mu(\omega) + j\omega Cu(\omega) + Ku(\omega) = F(\omega) \tag{J.1b}$$

Using a equivalent stiffness Matrix $H_{\mathrm{struct}}(\omega)$ solving (J.1b), under the assumption of constant matrices M, C and K, the equation can be expressed, shown in (J.2), with the displacement $u(\omega)$.

$$H_{\mathrm{struct}}^{-1}(\omega)u(\omega) = F(\omega) \tag{J.2}$$

Therefore the displacement of the structural system can be defined, using a transfer matrix H_{struct}, as

$$u(\omega) = H_{\text{struct}}(\omega)F(\omega). \tag{J.3}$$

Using the complex relations of velocity to the displacement in (J.4a) and the relation to the acceleration in (J.4b), the surface velocity is directly calculable in (J.5a), as well as the surface acceleration (J.5b).

$$v(t) = \dot{u}(t) \circ\!\!-\!\!\bullet\ v(\omega) = j\omega u(\omega) \tag{J.4a}$$

$$a(t) = \dot{v}(t) = \ddot{u}(t) \circ\!\!-\!\!\bullet\ a(\omega) = j\omega v(\omega) = -\omega^2 u(\omega) \tag{J.4b}$$

$$v(\omega) = j\omega H_{\text{struct}}(\omega)F(\omega). \tag{J.5a}$$

$$a(\omega) = -\omega^2 H_{\text{struct}}(\omega)F(\omega). \tag{J.5b}$$

K Modal Superposition

The principle of modal superposition is used in order to model the vibration as a set of individual modes. This approach can be seen as a summation of infinite individual structural modes, reproducing the physical behavior of the mechanical structure. In a FEM-sense the summation is limited by the total number of degrees of freedom N as,

$$u(t) = \sum_{}^{N} \phi_{\text{m}} y_{\text{m}}(t) \tag{K.1}$$

with the vibration eigenvectors ϕ_{m} and the modal coordinates $y_{\text{m}}(t)$. The vibration can be approximated with a fewer number of eigenforms M ($M < N$), as shown in (K.2).

$$u(t) = \sum_{\text{m}=0}^{M} \phi_{\text{m}} y_{\text{m}}(t) \tag{K.2}$$

Rewriting (K.2) in matrix expression according to $u = \phi y$ with $N \times M$ matrix $\phi = [\phi_1 \ldots \phi_M]$ and a vector $y(t)$ with length M, inserted in (J.1a), leads to the expression

$$\phi^{\text{T}} M \phi \ddot{y}(t) + \phi^{\text{T}} C \phi \dot{y}(t) + \phi^{\text{T}} K \phi y(t) = \phi^{\text{T}} F(t) \tag{K.3}$$

where,

$$\phi^{\text{T}} M \phi =: m,\ \phi^{\text{T}} C \phi =: c,\ \phi^{\text{T}} K \phi =: k,\ \phi^{\text{T}} F(t) =: f(t). \tag{K.4}$$

In case of entries of m, c and k on the main diagonal, the system is decoupled and can be solved in individual solutions for the m-th mode, as

$$m_{\text{m}} \ddot{y}(t) + c_{\text{m}} \dot{y}(t) + k_{\text{m}} y(t) = f_{\text{m}}(t). \tag{K.5}$$

The system is now decoupled in the modal coordinates using the mass m_m, damping c_m and stiffness k_m for the m-th mode, as shown in (K.6).

$$k_\mathrm{m} = \omega_{0,\mathrm{m}}^2 m_\mathrm{m} \text{ and } c_\mathrm{m} = 2\xi_\mathrm{m}\omega_{0,\mathrm{m}} m_\mathrm{m} \tag{K.6}$$

The mass-spring-damper system is reformulated in the TTD, as shown in (K.7) and in the TFD, as shown in (K.8).

$$\ddot{y}(t) + 2\xi_\mathrm{m}\omega_{0,\mathrm{m}}\dot{y}(t) + \omega_{0,\mathrm{m}}^2 y(t) = \frac{f_\mathrm{m}(t)}{m_\mathrm{m}} \tag{K.7}$$

$$(-\omega^2 + \mathrm{j}\omega 2\xi_\mathrm{m}\omega_{0,\mathrm{m}} + \omega_{0,\mathrm{m}}^2)y(\omega) = \frac{f_\mathrm{m}(\omega)}{m_\mathrm{m}} \tag{K.8}$$

In order to use the transfer matrix $H_\mathrm{struct}(\omega)$ from (J.3), the surface displacement is expressed as

$$u(\omega) = \sum_{m=0}^{M} H_\mathrm{struct,m}^{-1}(\omega)f_\mathrm{m}(\omega) \tag{K.9}$$

where the transfer matrix $H_\mathrm{struct}(\omega)$ is defined as shown in (K.10).

$$H_\mathrm{struct,m}(\omega) = (\omega_{0,\mathrm{m}}^2 + \mathrm{j}\omega 2\xi_\mathrm{m}\omega_{0,\mathrm{m}} - \omega^2)m_\mathrm{m} \tag{K.10}$$

The same procedure is performable for the surface velocity, as well as the surface acceleration, as

$$v(\omega) = \mathrm{j}\omega \sum_{m=0}^{M} H_\mathrm{struct,m}(\omega)f(\omega). \quad \text{(K.11a)} \qquad a(\omega) = -\omega^2 \sum_{m=0}^{M} H_\mathrm{struct,m}(\omega)f(\omega). \quad \text{(K.11b)}$$

Damping

The mechanical damping factor ξ_m of the m-th mode can be calculated in three different ways.

- Constant damping: The damping parameter is rated beneath $\xi = 0 \dots 1$ to be the same for all modes.

- Mode dependent constant damping: The damping $\xi_\mathrm{m} = 0 \dots 1$ is constant for the m-th mode, but can be different for each mode.

- Mode dependent variable damping: The damping of the m-th mode increases in dependency of the eigenfrequency or force.

The first approach is the most simplest, but has the disadvantage when considering higher modes. Normally at higher modes the damping is increased. This behavior is used in the second approach for mode dependent constant damping. The Last approach with a variable damping relating to the eigenfrequency or force is less used, because of the possibility of an increasing error. However, the damping is also influenced by the temperature of the system, which aggravates the definition of the damping factor.

Mass

The mass of the system is constant for the system and definable by the materials' mass density. Using mass-normalization for the eigen mode shapes, the mass matrix in the modal space is represented by the standard unity matrix.

Stiffness

The stiffness for linear materials is mostly modeled with the Hooke's law strain tensor ϵ, stress tensor σ and the elasticity tensor E, as shown in (K.12).

$$\epsilon = \frac{1}{E}\sigma \tag{K.12}$$

L Force Response Calculation

This section relates to the possibility of creating the structural transfer function with an structural FEM simulation, using ANSYS Mechanical [Inc17a] or Simulia Abaqus [Sys17]. Both transfer function can be used in section 4.4 in order to synthesize the surface displacement [Boe13; Roi09; vdGie11].

The j-th force excitation $F_j(\omega)$ is defined as the spatial force distribution $\mathcal{F}_j$ and the temporal force amplitude $\underline{\hat{F}}_j(\omega)$, as shown in (L.1).

$$F_j(\omega) = \mathcal{F}_j\underline{\hat{F}}_j(\omega) \tag{L.1}$$

The spatial force distribution $\mathcal{F}_j$ of harmonic response analysis is performed in a FEM simulation using a spatial tooth force shape of mode j, shown in section 4.2, with the normalized temporal force amplitude $\hat{F}_j(\omega)$ for all temporal frequencies $\omega = 2\pi f$. With the modal force vector $f = \phi^{\mathrm{T}}F$ from (K.5), the application leads to displacement response u, as (L.2), creating a normalized surface displacement transfer function $H_{u_{\mathrm{norm,j}}}(\omega)$, including the modal shapes m.

$$u_j(\omega) = H_{u_{\mathrm{norm,j}}}(\omega)\hat{\underline{F}}_j(\omega), \text{ with } H_{u_{\mathrm{norm,j}}}(\omega) = \sum_{m=0}^{M} \phi_m y_{m,j}(\omega) \tag{L.2}$$

The transfer function $H_{u_{\mathrm{norm,j}}}(\omega)$ links the force excitation $F(\omega)$ to the surface displacement $u_j(\omega)$, where the maximum of normalized temporal force amplitude $\max(\hat{\underline{F}}_j(\omega)) = 1N$.

The spatial forces can be applied in the structural FEM simulation in two different approaches, leading to the same results:

1. Standing waves: This approach describes two spatially standing orthogonal waves, changing their temporal force amplitude $\hat{F}_j$ for the j-th force amplitude in dependency of the frequency ω, as

$$F_j(\omega) = \hat{F}_j\bigg|_{\hat{F}_j=1N} e^{j\omega t}(cos(\alpha j) + sin(\alpha j)). \tag{L.3}$$

2. Rotating waves: Represents a two spatial waves, changing their phase angle with rotating in positive and negative direction, as

$$F_j(\omega) = \hat{F}_j\bigg|_{\hat{F}_j=1N} e^{j(\alpha j \pm \omega t)}. \tag{L.4}$$

M Extension for Sound Power Calculation

This section relates to the sound power calculation and subsequently the audible noise given from the induction machine to the environment. Calculating audible noises can have artifacts and measurements need special measurement environment and equipment, that can be costly and not purposeful.

Radiated sound power

The radiated sound power P from the machine surface is defined in (M.1), where ρ_0 and c_0 are the density and sound speed in the ambient medium, as well as the surface radiating area S and the radiation efficiency σ. The surface area can be estimated in summation of the surface segments S_i, exemplary relying on the finite-element mesh discretization [GWC06].

$$P = \sigma\rho_0 c_0 S|\bar{v}_n|^2, \text{ where } S = \sum_{i=1}^{I} S_i \text{ and } |\bar{v}_n| = \frac{1}{I}\sum_{i=1}^{I} \bar{v}_{i,n} \tag{M.1}$$

The radiated sound power is a quantity and position independent. In case of a test bench measurement, the sound power level can show dependencies on the position of the microphones. This procedure is relativized by taking the mean of multiple microphones in a test bench measurement [Sta10].

Radiation efficiency

The calculation of the radiation efficiency shows a high dependency on the geometric shape of the surface structure. As there are diverse variations of it, the radiation calculation can show huge differences. In general, the radiation efficiency σ is defined as the ratio of the radiated sound power P to the free-field sound power P_0, as shown in (M.2), and gives a relation between the radiated sound power P of the structure and the total vibrational energy E_v [GWC06; TL94; vdGie11; Bes08].

$$\sigma = \frac{P}{P_0} \propto \frac{P}{E_\mathrm{v}} \tag{M.2}$$

A rough estimation of induction machines' radiation efficiency can be performed using an analytical expression for a cylindrical radiator, as

$$\sigma_\mathrm{j}(k_0 a) = \frac{(k_0 a)^2 [Y_\mathrm{j}(k_0 a) J_{\mathrm{j}+1}(k_0 a) - J_\mathrm{j}(k_0 a) Y_{\mathrm{j}+1}(k_0 a)]}{[\mathrm{j} J_\mathrm{j}(k_0 a) - k_0 a J_{\mathrm{j}+1}(k_0 a)]^2 + [\mathrm{j} Y_\mathrm{j}(k_0 a) - k_0 a Y_{\mathrm{j}+1}(k_0 a)]^2}, \tag{M.3}$$

where the J_j and $J_{\mathrm{j}+1}$ are the first kind Bessel functions of the j-th and (j + 1)-th force and the Y_j and $Y_{\mathrm{j}+1}$ are the second kind Bessel functions of the j-th and (j + 1)-th force, as well as $k_0 = {}^\omega\!/_{c_0}$ and a as the stator outer radius.

With increasing structural vibration frequency, the radiation efficiency becomes 1, as

$$\lim_{\omega \to \infty} \sigma = 1, \tag{M.4}$$

and can be assumed as $\sigma \approx 1$ for a first approximation.

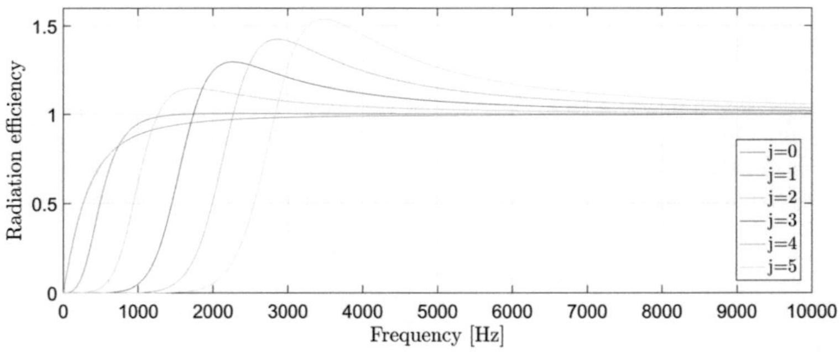

Figure M.1: Radiation efficiency factor of an cylindrical radiator for j=0,1,2,...5 (Based on [GWC06])

FEM and Boundary Element method (BEM) based acoustic simulation

For acoustic simulations of free-field sound, there are two possibilities of performing more accurate results beneath analytical expressions. In general, the most convenient methods are differed into FEM based calculations, using a perfectly matched layer PML and BEM based calculations. On the one hand, the PML method uses a artificial absorbing layer for wave equations, in which sound is not reflected back into the FEM [BT09; Cos89]. On the other hand, the BEM simulation only solves the boundary condition and with it only needs to be discretized on its the boundary. Both simulation methods can be used in order to simulate the radiated sound power and the sound pressure given to the environment.

With these simulation methods, a acoustic transfer vector (ATV) can be created in order to set the surface velocity of the induction machine in relation to the sound pressure at field points in the environment [Ger+01 MTG01], as

$$p(\omega) = \mathrm{ATV}^{\mathrm{T}}(\omega)v_{\mathrm{n}}(\omega). \tag{M.5}$$

The ATV can also be used in the modal structure decomposition, for each structural mode m, using a normalized vibration excitation in order to create a velocity transfer matrix $H_{\mathrm{Pnorm,j}}(\omega)$ for the j-th force shape excited surface velocity, as

$$H_{\mathrm{Pnorm,j}}(\omega) = \sum_{m=0}^{M} \mathrm{ATV}_{\mathrm{m}}^{\mathrm{T}}(\omega)\phi_{\mathrm{m}}y_{\mathrm{m,j}}(\omega), \tag{M.6}$$

synthesizing the sound pressure $p(\omega)$, including J force shapes, as

$$p(\omega) = \sum_{j=0}^{J} H_{\mathrm{Pnorm,j}}(\omega)\hat{F}_{\mathrm{j}}(\omega). \tag{M.7}$$

This procedure is also performable with various analytical models, as shown in the previous sections, which can be accurate alternatives. However, analytical expressions need a lot of knowledge about the system itself and can lead to huge differences, if the models are not validated step by step.

List of Figures

1.1 Construction of an Electrical Vehicle (EV) 1
1.2 Statistics of electrical vehicles 2
1.3 Example for priority of design characteristics of multi-objective optimization 3
1.4 Causality chain from voltage excitation to acoustic noises 4
1.5 Simulation path for efficient NVH simulations (based on [Kot+17b; Boe13]) . 5

2.1 Noise emission of automobiles (based on [Kön07]) 8
2.2 Noise sources in electrical drives 9
2.3 Transformation from TTD and STD to TFD and SFD 13
2.4 FT from TTD or STD to TFD or SFD 14
2.5 FFT based spectrum from TTD to TFD 14
2.6 Bode diagram of (a) HPF and (b) LPF 16
2.7 Bode diagram of a) NF and b) PF 17
2.8 Schematic diagram of a cascade controlled system 18
2.9 3-phase (a) inverter with load and (b) operation vectors 19
2.10 Switching sequence for SVPWM 19
2.11 SVPWM (a) switching and duty cycle and (b) harmonic distribution example 20
2.12 Switching sequence for FTPWM using (a) V_0 and (b) V_7 as zero vector . . . 21
2.13 FTPWM (a) switching and duty cycle and (b) harmonic distribution example 22
2.14 OVPWM (a) switching and duty cycle and (b) harmonic distribution example 22

3.1 System integration of the sequential electromagnetic model 25
3.2 Equivalent circuit of the fundamental flux model 26
3.3 Fundamental flux model (a) d-axis component and (b) q-axis component . . 27
3.4 Schematic diagram of the ideal rotating transformer (IRTF) 28
3.5 IRTF fundamental flux model (based on [DPV11]) 28
3.6 Fundamental flux model with rotor harmonics extension 29
3.7 IRTF fundamental flux model with rotor harmonics extension 30
3.8 Stator dependent rotor current harmonics for mean fundamental (a) d-axis
    and (b) q-axis stator current 30
3.9 Stator winding scheme for one pole pair 31
3.10 Stator (a) electric loading and (b) mmf 31
3.11 Stator winding function for U-Phase 33
3.12 Comparison of stator mmf calculations in (a) STD and (b) SFD 33
3.13 Run-up of stator mmf simulations in (a) STD and (b) SFD 34

3.14 Rotor (a) bar electric loading and (b) electric loading, mmf 35
3.15 Rotor winding function for two (a) neighbored and (b) pole pair repeating rotor bars 36
3.16 Run-up of rotor mmf simulation in (a) STD and (b) SFD 36
3.17 Comparison of rotor mmf calculations in (a) STD and (b) SFD 38
3.18 Spatial (a) stator and (b) rotor slot function 39
3.19 Air-gap permeance for closed rotor slots in (a) STD and (b) SFD 40
3.20 Air-gap permeance for stator and rotor opened slots in (a) STD and (b) SFD 41
3.21 Magneto motive force parts for saturation calculation 42
3.22 Schematic of the analytical saturation simulation 42
3.23 Saturated magnetic permeance in (a) STD and (b) SFD 43
3.24 Run-up of the magnetic permeance for (a) STD and (b) SFD calculation . . 44
3.25 FEM comparison of unsaturated constant air-gap flux density and saturated flux density 44
3.26 Coefficient estimation based on FEM magnetic flux density simulation 44
3.27 Comparison of mmf to material curve for saturation simulation 45
3.28 Material curve compared to mmf for saturation simulation 45
3.29 $B - \Theta$ material curve 46
3.30 Non linear saturation factor 46
3.31 Schematic of eccentric rotor position 46
3.32 Flux behavior for eccentric rotor position 46
3.33 Static eccentric air-gap permeance in a) STD and b) SFD 47
3.34 Schematic of axial segmentation 48
3.35 Spatial magnetic flux density in (a) STD and (b) SFD 50
3.36 Run-up of the magnetic flux density in (a) STD and (b) SFD 50
3.37 Run-up of the radial force in (a) STD and (b) SFD 52
3.38 System integration of the integrated electromagnetic model 53
3.39 Qualitative diagram of the origins and propagation of machine harmonics . . 53
3.40 Schematic of the modular air-gap flux approach 54
3.41 Stage 1: LILAG 55
3.42 Stage 2: LIDAG without saturation 58
3.43 Saturation factor for stator yoke and rotor core saturation 60
3.44 Stator phase currents for OP1 in (a) TTD and (b) TFD 65
3.45 Stator phase currents for OP2 in (a) TTD and (b) TFD 65
3.46 Rotor bar currents for OP1 in (a) TTD and (b) TFD 66
3.47 Rotor bar currents for OP2 in (a) TTD and (b) TFD 66
3.48 Electromagnetic torque for OP1 in (a) TTD and (b) TFD 67
3.49 Electromagnetic torque for OP2 in (a) TTD and (b) TFD 67
3.50 Temporal comparison of mmfs beneath one stator tooth for (a) OP1 and (b) OP2 in TTD 68
3.51 Spatial comparison of magnetic flux densities at one time step for (a) OP1 and (b) OP2 in STD 69

3.52 Spatial comparison of magnetic flux densities at one time step for (a) OP1 and (b) OP2 in SFD 70

3.53 Temporal comparison of magnetic flux densities beneath one stator tooth for (a) OP1 and (b) OP2 in TFD 70

3.54 Temporal comparison of magnetic flux densities beneath one stator tooth for (a) OP1 and (b) OP2 in TTD 71

3.55 Spatial comparison of radial forces at one time step for (a) OP1 and (b) OP2 in STD 72

3.56 Spatial comparison of radial forces at one time step for (a) OP1 and (b) OP2 in SFD 73

3.57 Temporal comparison of radial forces beneath one stator tooth for (a) OP1 and (b) OP2 in TFD 73

3.58 Temporal comparison of radial forces beneath one stator tooth for (a) OP1 and (b) OP2 in TTD 74

3.59 Stator phase currents for OP1 in (a) TTD and (b) TFD 75

3.60 Stator phase currents for OP2 in (a) TTD and (b) TFD 75

3.61 Rotor bar currents for OP1 in (a) TTD and (b) TFD 76

3.62 Rotor bar currents for OP2 in (a) TTD and (b) TFD 77

3.63 Electromagnetic torque for OP1 in (a) TTD and (b) TFD 78

3.64 Electromagnetic torque for OP2 in (a) TTD and (b) TFD 78

3.65 Spatial comparison of mmfs at one time step for (a) OP1 and (b) OP2 in STD 79

3.66 Spatial comparison of mmfs beneath one stator tooth for (a) OP1 and (b) OP2 in TTD 80

3.67 Spatial comparison of magnetic flux densities at one time step for (a) OP1 and (b) OP2 in STD 81

3.68 Spatial comparison of magnetic flux densities at one time step for (a) OP1 and (b) OP2 in SFD 82

3.69 Temporal comparison of magnetic flux densities beneath one stator tooth for (a) OP1 and (b) OP2 in TFD 82

3.70 Temporal comparison of magnetic flux densities beneath one stator tooth for (a) OP1 and (b) OP2 in TTD 83

3.71 Spatial comparison of radial forces at one time step for (a) OP1 and (b) OP2 in STD 84

3.72 Spatial comparison of radial forces at one time step for (a) OP1 and (b) OP2 in SFD 85

3.73 Temporal comparison of radial forces beneath one stator tooth for (a) OP1 and (b) OP2 in TFD 85

3.74 Temporal comparison of radial forces beneath one stator tooth for (a) OP1 and (b) OP2 in TTD 86

3.75 FT based spectrum of the stator current U-phase for (left) measured and (right) simulated speed run-up from 0 rpm to 18000rpm at 118Nm of desired torque 87

3.76 Comparison of torque for a run-up from 0 to 18000rpm at 118Nm of desired torque 87

3.77 (a) Upper end winding connection of the stator u-phase for a 48 slot stator 2 pole pair machine and (b) Mean line segment for the 4 windings with effective turns strength of 4 for c'd' and 2 for cc' and d'd 89

4.1 System integration of the structural dynamic model 91

4.2 Spatial decomposition of the radial force 93

4.3 Magnitude of the radial force (a) mode 0 and (b) mode 4 in TTD 93

4.4 Temporal decomposition of radial force (a) mode 0 and (b) mode 4 94

4.5 Stator tooth force projection (a) sine and (b) tooth / tsine 95

4.6 Comparison of (a) complete stator, (b) reduced stator and (c) stator teeth model. 99

4.7 Comparison of radial forces for (a) air-gap, (b) sine, (c) tooth and (d) tsine approach. 100

4.8 Comparison of TFs for the (a) 0th, (b) 4th, (c) 8th and (d) 12th force shape j.101

4.9 Resultant (a) 0th, (b) 4th and (c) 8th force shape at one time step. 102

4.10 Modal force matrix (MFM) for (a) real and (b) imaginary component 102

4.11 Resultant (a) breathing @ 8781Hz , (b) breathing @ 8694Hz, (c) breathing-bending @ 8930Hz, (d) bending @ 9200Hz, (e) axial-radial-contraction @ 9887Hz and (f) axial-radial-contraction @ 9897Hz structural mode shape. . . 103

4.12 Validation of (a) measured and (b) simulated FT based surface velocity spectrum. 104

4.13 Surface velocity level for measured and simulated speed run-ups 105

5.1 Schematic diagram of noise-optimized and efficiency-optimized operating points.107

5.2 Magneto motive force (mmf) for reference stator d-axis current at 60Nm of torque. 108

5.3 Magneto motive forces (mmfs) for (a) minimum and (b) maximum stator d-axis current at 60Nm of torque. 109

5.4 Schematic diagrams for (a) reference, (b) minimum and (c) maximum stator d-axis current at 60Nm of torque. 110

5.5 Radial force spectrum for (a) reference, (b) minimum and (c) maximum stator d-axis current at 60Nm of torque. 110

5.6 Radial force levels for speed run-ups at 60Nm of torque. 111

5.7 Surface velocity levels for speed run-ups at 20Nm of torque. 112

5.8 Surface velocity spectrum for (a) reference,(b) minimum and (c) maximum stator d-axis current at 20Nm of torque. 113

5.9 Surface velocity levels for speed run-ups at (a) 60Nm and (b) 100Nm of torque.113

5.10 Surface velocity spectrum for (a) reference, (b) minimum and (c) maximum stator d-axis current at 60Nm of torque. 114

5.11 Surface velocity spectrum for (a) reference, (b) minimum and (c) maximum stator d-axis current at 100Nm of torque. 114

5.12 Maximum of velocity level difference for different stator d-axis current at 20Nm of torque. 115

5.13 Maximum of velocity level difference for different stator d-axis current at (a) 60Nm and (b) 100Nm of torque. 115

5.14 Optimized stator d-currents changing with torque reference. 117

5.15 Optimized operating points at the fundamental speed range. 117

5.16 Stator d-axis current for noise optimized operating at (a) 20Nm, (b) 40Nm, (c) 60Nm, (d) 80Nm, (e) 100Nm and (f) 120Nm of torque. 118

5.17 Measured surface velocities with NVH- and Efficiency-optimized operating strategies. 119

5.18 Measured FT based stator current spectrum at 60Nm of torque for (a) reference, (b) minimum and (c) maximum flux 120

5.19 Simulated FT based rotor bar current spectrum at 60Nm of torque for (a) reference, (b) minimum and (c) maximum flux 121

5.20 Simulated FT based electromagnetic torque spectrum at 60Nm of torque for (a) reference, (b) minimum and (c) maximum flux 121

6.1 Schematic diagram of the harmonics elimination. 123

6.2 Signal flow chart of the open loop (a) coupled and (b) decoupled harmonic signal injection 124

6.3 Schematic diagram of eccentric rotor position 125

6.4 Harmonic Repetitive controller in d-q-axis coordinate system 127

6.5 Transfer behavior of repetitive controller (a) magnitude and (b) phase 127

6.6 Digital adaptive filter 128

6.7 Optimization method 128

6.8 Harmonic PI-controller in d-q-axis coordinate system 129

6.9 Transfer behavior of harmonic PI-controller (a) magnitude and (b) phase . . 129

6.10 Block diagram of (a) PR- and (b) VPI-controller in d-q-axis system 130

6.11 Transfer behavior of the PR- and VPI-controller (a) magnitude and (b) phase 131

6.12 Transfer behavior of the PR-controller without and with delay time compensation (a) magnitude and (b) phase 132

6.13 Transfer behavior of the VPI-controller without and with delay time compensation (a) First-order hold (FOH) magnitude, (b) First-order hold (FOH) phase, (c) Impulse invariant (IMP) magnitude and (d) Impulse invariant (IMP) phase. 134

6.14 Signal flow chart of the closed loop (a) coupled and (b) decoupled harmonic signal injection 135

6.15 Measurement with voltage injection at $\sim 10\%$ of eccentricity 137

6.16 Measurement with voltage injection at $\sim 40\% - 50\%$ of eccentricity 137

6.17 Measured FT based stator current spectrum of 6th harmonic elimination for (a) VPI 0Nm, (b) VPI 20Nm, (c) VPI 40Nm, (d) PR 0Nm, (e) PR 20Nm and (f) PR 40Nm of torque and 3000rpm of speed 139
6.18 Measured FT based surface velocity spectrum of 6th harmonic elimination with VPI controller for (a) 0Nm, (b) 20Nm and (c) 40Nm of torque and 3000rpm of speed 140
6.19 Measured FT based direct current spectrum of 6th harmonic elimination with (a) VPI and (b) PR controller for 20Nm of torque and 3000rpm of speed . . 141
6.20 Measured FT based spectrum of 17.1th harmonic elimination with VPI-controller for (a) stator current, (b) surface velocity, (c) direct current for 20Nm of torque and 1500rpm of speed 142
6.21 Measured FT based stator current spectra of the 6th harmonic in stator frame and 18th harmonic in rotor frame system elimination with (a) VPI and (b) PR controller for 20Nm of torque and 3000rpm of speed 144
6.22 Measured FT based stator current spectrum of 6th harmonic in stator frame and 18th harmonic in rotor frame system elimination for (a) VPI 20Nm, (b) VPI 40Nm, (c) VPI 60Nm, (d) PR 20Nm, (e) PR 40Nm and (f) PR 60Nm of torque and 1000rpm of speed 145
6.23 Measured FT based stator current spectra of speed run-ups (a) without, (b) VPI and (c) PR 6th harmonic elimination for 60Nm of torque from 250-5000rpm of speed 147

B.1 Signal flow chart of the controlled system 154
M.1 Radiation efficiency factor of an cylindrical radiator for j=0,1,2,...5 (Based on [GWC06]) 164

List of Tables

3.1   Operating points (OPs) for comparison 64

6.1   Relations of discretization methods (based on [Yep+10]) 132

C.1   Electromagnetic simulation methods (based on [ENG13]) 155
F.1   Spatial convolution of the air-gap permeance and saturation permeance orders 156
F.2   Spatial convolution of the resulting mmf and the saturated permeance orders 157
F.3   Spatial convolution of the magnetic flux density orders with itself 157
G.1   Overview of the compared simulation results 158

Nomenclature

ATV	Acoustic transfer vector
A_{surf}	Surface area
A_ν	Magnitude of complex Fourier coefficients
$A_\triangle$	Area of a thin strip on the curved surface
$B_{\mathrm{air-gap}}$	Radial magnetic flux density in the air-gap
B_{r}	Radial magnetic flux density
B_{t}	Tangential magnetic flux density
C	Damping matrix
C	Inverse leakage inductance matrix
$C_{\sigma,\mathrm{r}}$	Inverse leakage inductance matrix of rotor
$C_{\sigma,\mathrm{s}}$	Inverse leakage inductance matrix of stator
D	Ratio of sampling time to signal cycle duration
E	Elasticity tensor
F	Force
F	Function in frequency domain
F_{j}	Force of the j-th force mode
F_{r}	Radial force
$F_{\mathrm{r,cos}}$	Real coefficient of radial force mode
$F_{\mathrm{r,sin}}$	Imaginary coefficient of radial force mode
F_{tooth}	Stator tooth force
F_{tsine}	Stator tooth force with tsine approach
G	Linkage matrix
$G_{\mathrm{PI_h}}$	Transfer function of harmonic PI-controller
G_{PR}	Transfer function of PR-controller
G_{R}	Transfer function of repetitive controller
G_{VPI}	Transfer function of VPI-controller
G_{f}	Transfer function to guarantee stability of repetitive controller
H	Low-pass filter of repetitive controller
H_{HPF}	Transfer function of high-pass filter
H_{LPF}	Transfer function of low-pass filter

H_{NF}	Transfer function of notch filter
H_{PF}	Transfer function of peak filter
$H_{\mathrm{P_{norm,j}}}$	Velocity transfer matrix of j-th force shape
H_{struct}	Structural transfer matrix
$H_{\mathrm{struct,m}}$	Structural transfer matrix of the m-th structural mode
H_{tooth}	Transformation matrix for tooth approach
H_{tsine}	Transformation matrix for tsine approach
$H_{\mathrm{u_{norm,j}}}$	Normalized transfer function for j-th force mode
I_{d}	Stator d-axis current
$I_{\mathrm{d,eff}}$	Stator d-axis current in efficiency operating point
$I_{\mathrm{d,max}}$	Stator d-axis current in maximum flux operating point
$I_{\mathrm{d,min}}$	Stator d-axis current in minimum flux operating point
$I_{\mathrm{d,opt}}$	Stator d-axis current in optimum operating point
$I_{\mathrm{d,ref}}$	Stator d-axis current in reference flux operating point
J	Number of force shapes
K	Stiffness matrix
$K_{\mathrm{I_h}}^{\mathrm{PI}}$	Integral component of harmonic PI-controller
$K_{\mathrm{P_h}}^{\mathrm{PI}}$	Proportional component of harmonic PI-controller
$K_{\mathrm{I}}^{\mathrm{PR}}$	Integral component of PR-controller
$K_{\mathrm{P}}^{\mathrm{PR}}$	Proportional component of PR-controller
$K_{\mathrm{I}}^{\mathrm{VPI}}$	Integral component of VPI-controller
$K_{\mathrm{P}}^{\mathrm{VPI}}$	Proportional component of VPI-controller
K_{d}	Derivative component
K_{i}	Integral component
K_{p}	Proportional component
L_{M}	Main flux inductance in transformed system
L_{M}^{ν}	Main inductance in the transformed system of ν-th harmonic order
L_{m}	Main flux inductance in main system
$L_{\sigma,\mathrm{R}}$	Rotor leakage inductance in transformed system
$L_{\sigma,\mathrm{R}}^{\nu}$	Rotor leakage inductance in the transformed system of ν-th harmonic order
$L_{\sigma,\mathrm{S}}$	Stator leakage inductance in transformed system
$L_{\sigma,\mathrm{r}}$	Rotor leakage inductance in rotor system
$L_{\sigma,\mathrm{s}}$	Stator leakage inductance in stator system
M	Filter size
M	Magnetization matrix
M	Mass matrix

M	Structural eigenforms
MFM	Modal force matrix
N	Matrix of winding functions
N	Number of delay shifts
N	Number of samples
N	Structural degrees of freedom
$N_{\mathrm{el,r}}^{\mathrm{q_r}}$	Rotor electric load winding function of q_r-th rotor bar
$N_{\mathrm{el,s}}^{\mathrm{m}}$	Stator electric load winding function of m-th phase
N_{r}	Winding functions of rotor
N_{s}	Winding functions of stator
$N_{\Theta,\mathrm{r}}^{\mathrm{q_r}}$	Rotor mmf winding function of q_r-th rotor bar
$N_{\Theta,\mathrm{s}}^{\mathrm{m}}$	Stator mmf winding function of m-th phase
P	Period
P	Radiated sound power
Q_{r}	Number of rotor slots
Q_{s}	Number of stator slots
R_{R}	Rotor resistance in transformed system
R_{R}^{ν}	Rotor resistance in transformed system of ν-th harmonic order
R_{r}	Rotor resistance in rotor system
R_{s}	Stator resistance in stator system
S	Surface area of radiation
S_{i}	Segments of surface of radiated area
T_{C}	Clark Transformation
T_{P}	Park Transformation
T_{p}	Signal cycle duration
T_{s}	Sampling time
W	Delay component of repetitive controller
a	Factor for universal flux orientation
a	Real component of Fourier coefficient
a	Surface acceleration
a_{el}	Spatial electric stator load
$a_{\mathrm{el,r}}^{\mathrm{q_r}}$	Spatial electric rotor load of q_r-th rotor bar
b	Imaginary component of Fourier coefficient
$b_{\mathrm{q_r}}$	Rotor slot width
b_{s}	Stator slot width
c	Compex number of Fourier coefficients

c_0	Sound speed of ambient medium
c_m	Damping of the m-th structural mode
c_ratio	Tooth to slot ratio
e	Input error
e_r	Factor of entering of the magnetic flux density of rotor slot
e_s	Factor of entering of the magnetic flux density of stator slot
f	Function in time domain
f_OVPWM	Frequency components of OVPWM
f_SVPWM	Frequency components of SVPWM
f_c	Cut of frequency
f_ecc	Excited frequency for eccentric rotor position
f_elec	Electrical frequency
f_m	Force of the m-th structural mode
f_p	Frequency of sampling rate
f_r	Rotor slot function
f_s	Stator frequency
f_s	Stator slot function
$f_\mathrm{sat,r}$	Rotor saturation function
$f_\mathrm{sat,s}$	Stator saturation function
f_switch	Inverter switching frequency
g_fic	Fictive air-gap function
h	Temporal harmonic order
h_r	Induced temporal harmonic orders of rotor system
h_s	Induced temporal harmonic orders of rotor system
i_Mag	Magnetization current in transformed system
i_Mag^ν	Magnetization current of ν-th harmonic order
i_R	Rotor current in transformed system
i_R^xy	Rotor current in rotor system
$i_\mathrm{R}^{\mathrm{xy},\nu}$	Rotor current in rotor system of ν-th harmonic order
i_U	Stator current U-phase
i_V	Stator current V-phase
i_W	Stator current W-phase
i_d	Stator current d-axis component
$i_\mathrm{d,R}$	Rotor current d-axis component
i_q	Stator current q-axis component
i_r	Rotor current in rotor system

$i_r^{q_r}$	Rotor current of q_r-th rotor bar
i_r^{ν}	Rotor current of ν-th harmonic order
i_s	Stator current in stator system
i_s^m	Stator current of m-th phase
i_α	Stator current α-component
i_β	Stator current β-component
k	Spatial velocity
k_{HPF}	Filter coefficient of high-pass filter
k_{LPF}	Filter coefficient of low-pass filter
k_{NF}	Filter coefficient of peak filter
k_{PF}	Filter coefficient of notch filter
k_m	Stiffness of the m-th structural mode
$k_{sat,0}$	Saturation coefficient of 0-th saturation order
$k_{sat,2}$	Saturation coefficient of 2-th saturation order
$k_{sat,r}$	Rotor saturation scaling factor
$k_{sat,s}$	Stator saturation scaling factor
k_{skw}	Skewing part
l_{Fe}	Axial length of active parts of the induction machine
m	Stator phase
m_m	Mass of the m-th structural mode
m_s	Total number of stator phase
n	Temporal harmonic order
n_{ecc}	Factor of eccentric rotor position
n_r	Number of conductors in rotor slot
n_s	Number of conductors in stator slot
n_{skw}	Total number skewing part
p	Pole pair number
p	Sound pressure
q_r	Rotor bar
r	Linked radius
r_{dyn}	Radius of static dynamic rotor position
r_{stat}	Radius of static eccentric rotor position
s	Induction machine slip
s	Laplace transformed
s_ν	Induction machine slip of ν-th harmonic order
t	Time

Nomenclature

tq	Electromagnetic torque
u	Force mode
u	Spatial frequency domain
u	Structural displacement
u_U	Stator voltage U-phase
u_V	Stator voltage V-phase
u_W	Stator voltage W-phase
u_d	Stator voltage d-axis component
u_q	Stator voltage q-axis component
u_s	Stator voltage in stator system
u_α	Stator voltage α-component
u_β	Stator voltage β-component
v	Surface velocity
v_n	Normal surface velocity
v_r	Radial surface velocity
w	Weighting factor
x	Injection signal
y	Function
$\hat{y}$	Amplitude
y_m	Modal coordinates of m-th harmonic order
y_n	Oscillation
$y_{\mathrm{n},\nu}$	Wave
z	Z transformed
Θ_Rj	Local mmf relating to rotor yoke
Θ_Rz	Local mmf relating to rotor tooth
Θ_Sj	Local mmf relating to stator yoke
Θ_Sz	Local mmf relating to stator tooth
Θ_g	Local mmf relating to air-gap
Θ_r	Rotor magneto motive force
Θ_s	Stator magneto motive force
$\Lambda_\mathrm{air-gap}$	Air-gap permeance
$\Lambda_{\mathrm{air-gap},\mathrm{k_{skw}}}$	Air-gap permeance of skewing part
Λ_sat	Saturation of air-gap permeance
Φ_R	Vector of modal response degree of freedom
$\Phi_\mathrm{m}^\mathrm{m}$	Main flux
Ψ_M	Main flux linkage in transformed system

Nomenclature

$\Psi_{\mathrm{M}}^{\mathrm{xy}}$	Main flux linkage in rotor system
$\Psi_{\mathrm{M}}^{\mathrm{xy},\nu}$	Main flux linkage in the rotor system of ν-th harmonic order
Ψ_{M}^{ν}	Main flux linkage of ν-th harmonic order
Ψ_{R}	Rotor flux linkage in transformed system
$\Psi_{\mathrm{R}}^{\mathrm{xy}}$	Rotor flux linkage in rotor system
$\Psi_{\mathrm{R}}^{\mathrm{xy},\nu}$	Rotor flux linkage in the rotor system of ν-th harmonic order
Ψ_{R}^{ν}	Rotor flux linkage of ν-th harmonic order
Ψ_{S}	Stator flux linkage in transformed system
α	Spatial points
δ	Mechanical air-gap width
ϵ	Strain tensor
η	Spectrum matrix
$\angle\theta_{\mathrm{r}}$	Mechanical rotor angle
θ_{R}	Rotor flux linkage angle
$\lambda_{\mathrm{sat,r}}$	Saturated permeance matrix of rotor system
$\lambda_{\mathrm{sat,s}}$	Saturated permeance matrix of stator system
μ	Spatial rotor order
μ	Step size
μ_0	Permeability of vacuum
μ_{i}	Permeability of iron
μ_{slot}	Rotor slotting harmonic order
μ_{winding}	Rotor winding harmonic order
μ^*	Rotor harmonic order without interaction stator slot harmonic orders
μ'	Spatial rotor harmonic order including saturation harmonic orders
$\tilde{\mu}$	Rotor harmonic order, pole pair relating
ν	Spatial harmonic order
ν	Spatial stator harmonic order
ν_{int}^*	Stator harmonic order in interaction with rotor slot harmonic orders
ν_{slot}	Stator slotting harmonic order
ν_{winding}	Stator winding harmonic order
ν^*	Stator harmonic order without interaction rotor slot harmonic orders
ν'	Spatial stator harmonic order including saturation harmonic orders
$\tilde{\nu}$	Stator harmonic order, pole pair relating
ξ	Error plane
ξ_{m}	Damping factor of the m-th eigenfrequency
$\xi_{\mathrm{r},\mu}$	Rotor winding factor of μ-th harmonic order

$\xi_{\mathrm{s},\nu}$	Stator winding factor of ν-th harmonic order
ρ_0	Density of ambient medium
σ	Harmonic order of repetitive controller
σ	Radiation efficiency
σ	Stress tensor
σ_{r}	Maxwell stress tensor
τ_{r}	Rotor temporal order
τ_{p}	Pole division radian
τ_{r}^*	Temporal harmonic orders of rotor system
τ_{s}^*	Temporal harmonic orders of stator system
τ_{tooth}	Tooth radian
τ_{tq}^*	Temporal harmonic orders of electromagnetic torque
φ	Angle
φ_{skw}	Skewing angle
ϕ_{m}	Mode shape of the m-th vibration
ψ_{m}	Main flux linkage
$\psi_{\mathrm{m}}^{\mathrm{m}}$	Main flux linkage of the m-th stator coil
$\psi_{\mathrm{m}}^{\mathrm{q_r},xy}$	Main flux linkage of the q_{r}-th rotor bar
ψ_{r}	Rotor flux linkage
ψ_{s}	Stator flux linkage
ω	Angular frequency
ω	Temporal frequency domain
ω_0	Fundamental angular frequency
$\omega_{0,\mathrm{m}}$	N-th eiegnfreqeuency of the m-th vibration
ω_{bw}	Angular band width frequency
ω_{c}	Angular cut of frequency
ω_{m}	Induction machine rotor angular frequency
ω_{p}	Angular frequency of sampling rate
ω_{s}	Stator angular frequency
ω_{slip}	Angular slip frequency

Acronyms

1-D	1-dimensional
2-D	2-dimensional
3-D	3-dimensional
AC	alternating current
APF	active power filter
ATV	acoustic transfer vector
BEM	Boundary Element method
DC	direct current
EF	Euler forward
EU	European Union
EV	Electrical Vehicle
FEM	Finite Element method
FFT	fast Fourier transform
FIR	finite impulse response
FOH	First-order hold
FT	Fourier transform
FTPWM	Flat top pulse width modulation
HPF	high-pass filter
IGBT	insulated-gate bipolar transistor
IIR	infinite impulse response
IMP	Impulse invariant
IRTF	ideal rotating transformer
LILAG	Leakage inclusive lumped air-gap model

LIDAG	Leakage inclusive distributed air-gap model
LMS	least-mean-square
LPF	low-pass filter
LUT	look-up-table
MATLAB	MATrix LABobratory
MFM	modal force matrix
mmf	magneto motive force
NF	notch filter
NVH	Noise, Vibration and Harshness
OP	operating point
OVPWM	Over pulse width modulation
P	proportional
PF	peak filter
PI	proportional-integral
PID	proportional-integral-derivative
PML	Perfectly matched layer
PR	proportional-resonant
PSM	permanently excited synchronous machine
PWM	pulse width modulation
SFD	spatial frequency domain
STD	spatial time domain
SVPWM	Space vector pulse width modulation
TF	transfer function
TFD	temporal frequency domain
THD	Total Harmonic Distortion
TP	Tustin with pre-warping
TTD	temporal time domain

US	United States
VPI	vector proportional-integral

Bibliography

[And15]   A. Andersson. "Electric machine control for energy efficient electric drive systems". PhD thesis. Chalmers University of Technology Goeteburg, Department of Energy and Environment, 2015 (cit. on p. 116).

[Ang+11] A. Angermann, M. Beuschel, M. Rau, and U. Wohlfarth. *MATLAB - Simulink - Stateflow*. Oldenbourg Wissenschaftsverlag GmbH, 2011 (cit. on p. 23).

[BTM14]   T. W. Basseet, S. Tate, and M. Maunder. "Study of high frequency noise from electric machines in hybrid and electric vehicles". In: *inter.noise* (Nov. 2014). URL: `https://www.acoustics.asn.au/conference_proceedings/%20INTERNOISE2014/papers/p578.pdf` (cit. on p. 8).

[BT09]   H. Beriot and M. Tournour. *On the locally-conformal perfectly matched layer implementation for Helmholtz equation*. Jan. 2009 (cit. on p. 165).

[Bes08]   J. L. Besnerais. "Reduction of magnetic noise in PWM-supplied induction machines". PhD thesis. Laboratiore d'Electricite et d'Electronique de Puissance de Lille Ecole Centrale de Lille, 2008 (cit. on pp. 23, 24, 31, 34, 38, 41, 42, 49, 51, 52, 92, 107, 108, 164).

[Bin12]   A. Binder. *Elektrische Maschinen und Antriebe*. 18th ed. Springer-Verlag Berlin Heidelberg, 2012 (cit. on pp. 24–26, 28, 29).

[Bis+16] W. M. Bischof, B. Chatterjee, M. Boesing, M. Hennen, and R. M. Kennel. "Modeling Radial Air-Gap Forces of Three-Phase Cage Induction Machines in Spatial Frequency Domain". In: *Electrical Machines and Systems (ICEMS), 2016 19th International Conference on*. Nov. 2016 (cit. on pp. 31, 32, 34, 35, 37, 38, 42, 46, 49, 51, 51, 107, 108, 110, 111).

[Bis+17] W. M. Bischof, B. Chatterjee, M. Boesing, M. Hennen, and R. M. Kennel. "Modeling Inverter-Fed Three-Phase Squirrel-Cage Induction Machines including Spatial and Temporal Harmonics". In: *International Electric Machines and Drives Conference 2017 (IEMDC 2017), Miami*. 2017 (cit. on pp. 45, 49, 51, 61, 107, 108, 110, 111).

[BHK15]   W. Bischof, M. Hennen, and R. Kennel. "Synthesized magnetic flux density in three-phase cage induction machines with MATLAB Simulink". In: *Electrical Machines and Systems (ICEMS), 2015 18th International Conference on*. Oct. 2015, pp. 109–115. DOI: `10.1109/ICEMS.2015.7385009` (cit. on pp. 31, 34, 38, 49, 51, 52, 107).

[Bis+01]   D. Bispo, L. Martins, J. de Resende, and D. de Andrade. "A new strategy for induction machine modeling taking into account the magnetic saturation". In: *Industry Applications, IEEE Transactions on* 37.6 (Nov. 2001), pp. 1710–1719. ISSN: 0093-9994. DOI: 10.1109/28.968182 (cit. on p. 52).

[Boe+12]   M. Boesing, M. Niessen, T. Lange, and R. De Doncker. "Modeling spatial harmonics and switching frequencies in PM synchronous machines and their electromagnetic forces". In: *Electrical Machines (ICEM), 2012 XXth International Conference on.* 2012, pp. 3001–3007. DOI: 10.1109/ICElMach.2012.6350315 (cit. on pp. 54, 94).

[Boe+10]   M. Boesing, T. Schoenen, K. A. Kasper, and R. W. D. Doncker. "Vibration Synthesis for Electrical Machines Based on Force Response Superposition". In: *IEEE Transactions on Magnetics* 46.8 (Aug. 2010), pp. 2986–2989. ISSN: 0018-9464. DOI: 10.1109/TMAG.2010.2042291 (cit. on pp. 92, 94).

[Boe13]   M. Boesing. "Acoustic Modeling of Eletrical Drives". PhD thesis. RWTH Aachen, 2013. URL: http://publications.rwth-aachen.de/record/459444/files/5201.pdf (cit. on pp. 4, 5, 8, 10, 92, 94–98, 111, 112, 158, 162).

[Boj+05]   R. Bojoi, G. Griva, V. Bostan, M. Guerriero, F. Farina, and F. Profumo. "Current control strategy for power conditioners using sinusoidal signal integrators in synchronous reference frame". In: *Power Electronics, IEEE Transactions on* 20.6 (Nov. 2005), pp. 1402–1412. ISSN: 0885-8993. DOI: 10.1109/TPEL.2005.857558 (cit. on p. 130).

[BN09]   I. Boldea and S. A. Nasar. *The Induction Machine Design Handbook.* Ed. by S. edition. CRC PRess, 2009 (cit. on pp. 24, 38, 45).

[Bos+04]   G. Bossio, C. De Angelo, J. Solsona, G. Garcia, and M. Valla. "A 2-D model of the induction machine: an extension of the modified winding function approach". In: *Energy Conversion, IEEE Transactions on* 19.1 (Mar. 2004), pp. 144–150. ISSN: 0885-8969. DOI: 10.1109/TEC.2003.822294 (cit. on p. 34).

[Bus15]   R. Busch. *Elektrotechnik und Elektronik.* Springer Fachmedien Wiesbaden, 2015 (cit. on p. 18).

[Cam24]   W. Campbell. "Protection of Steam Turbine Disk Whells from Axial Vibration". In: *Transactions of the American Society of Mechanical Engineers* (1924) (cit. on p. 13).

[Cha+96]   A. Charette, J. Xu, A. Lakhsasi, Z. Yao, and V. Rajagopalan. "Modeling and simulation of saturated induction motors". In: *Computers in Power Electronics, 1996., IEEE Workshop on.* Aug. 1996, pp. 163–167. DOI: 10.1109/CIPE.1996.612354 (cit. on p. 41).

[CZQ13]   D. Chen, J. Zhang, and Z. Qian. "Research on fast transient and 6n +- 1 harmonics suppressing repetitive control scheme for three-phase grid-connected inverters". In: *Power Electronics, IET* 6.3 (Mar. 2013), pp. 601–610. ISSN: 1755-4535. DOI: 10.1049/iet-pel.2012.0348 (cit. on p. 126).

[Che+07]   S. Chen, Y. Lai, S.-C. Tan, and C. Tse. "Optimal design of repetitive controller for harmonic elimination in PWM voltage source inverters". In: *Telecommunications Energy Conference, 2007. INTELEC 2007. 29th International.* Sept. 2007, pp. 236–241. DOI: 10.1109/INTLEC.2007.4448774 (cit. on p. 126).

[CS16]   H. Chuan and J. K. E. Shek. "Reducing Unbalanced Magnetic Pull of an induction machine through active control". In: *8th IET International Conference on Power Electronics, Machines and Drives (PEMD 2016).* Apr. 2016, pp. 1–6. DOI: 10.1049/cp.2016.0330 (cit. on p. 125).

[Com10]   T. M. Company. "TMC to Sell Approaching Vehicle Audible System for 'Prius'". In: *News Release* (2010). URL: http://www2.toyota.co.jp/en/news/10/08/0824.html (cit. on p. 7).

[Cor15]   M. Corporation. *Basic Total Harmonic Distortion (THD) Measurement.* Jan. 2015. URL: https://www.microsemi.com/document-portal/doc_view/134813-an30-basic-total-harmonic-distortion-thd-measurement (cit. on p. 14).

[Cos89]   M. Costabel. "Principles of Boundary Element Methods". MA thesis. Technische Hochschule Darmstadt, 1989 (cit. on p. 165).

[DPV11]   R. De Doncker, D. W. Pulle, and A. Veltman. *Advanced Electrical Drives - Analysis, Modeling, Control.* 1st ed. Springer Science+Business Media B.V., 2011, 462 p. (Cit. on pp. 17, 18, 26–29, 54, 55, 116).

[De 92]   R. De Doncker. "Field-oriented controllers with rotor deep bar compensation circuits [induction machines]". In: *Industry Applications, IEEE Transactions on* 28.5 (Sept. 1992), pp. 1062–1071. ISSN: 0093-9994. DOI: 10.1109/28.158830 (cit. on p. 23).

[Des+16]   G. Despret, M. Hecquet, V. Lanfranchi, and M. Fakam. "Skew effect on the radial pressure of induction motor". In: *2016 Eleventh International Conference on Ecological Vehicles and Renewable Energies (EVER).* Apr. 2016, pp. 1–6. DOI: 10.1109/EVER.2016.7476383 (cit. on p. 48).

[Don+99]   V. Donescu, A. Charette, Z. Yao, and V. Rajagopalan. "Modeling and simulation of saturated induction motors in phase quantities". In: *Energy Conversion, IEEE Transactions on* (1999). ISSN: 0885-8969 (cit. on p. 41).

[Dos88]   O. Dossing. *Structural Testing - Part II - Modal Analysis and Simulation.* Bruel and Kjar, 1988 (cit. on p. 96).

[Ede+07]   J. D. Ede, K. Atallah, G. W. Jewell, J. B. Wang, and D. Howe. "Effect of Axial Segmentation of Permanent Magnets on Rotor Loss in Modular Permanent-Magnet Brushless Machines". In: *IEEE Transactions on Industry Applications* 43.5 (Sept. 2007), pp. 1207–1213. ISSN: 0093-9994. DOI: 10.1109/TIA.2007.904397 (cit. on p. 48).

[Eff00]   N. Efford. *Digital Image Processing: A Practical Introduction Using Java*. Pearson Education, 2000 (cit. on p. 11).

[Ele12]   Elektroautor. "32 Vorteile von Elektroautos und täglich werden es mehr". In: *Elektroautor.vom* (2012). URL: http://www.elektroautor.com/die-vielen-vorteile-eines-elektroautos-wer-bietet-mehr/ (cit. on p. 2).

[ERD09]   S. Engel, K. Rigbers, and R. De Doncker. "Digital repetitive control of a three-phase flat-top-modulated grid tie solar inverter". In: *Power Electronics and Applications, 2009. EPE '09. 13th European Conference on*. Sept. 2009, pp. 1–10 (cit. on p. 126).

[ENG13]   E. ENGINEERING. *Manatee presentation*. 2013. URL: http://eomys.com/IMG/pdf/2017_06_14_manatee_presentation_en_light.pdf (cit. on p. 155).

[Esc+08]   G. Escobar, P. Hernandez-Briones, P. Martinez, M. Hernandez-Gomez, and R. Torres-Olguin. "A Repetitive-Based Controller for the Compensation of 6l +- 1 Harmonic Components". In: *Industrial Electronics, IEEE Transactions on* 55.8 (Aug. 2008), pp. 3150–3158. ISSN: 0278-0046. DOI: 10.1109/TIE.2008.921200 (cit. on p. 126).

[FSD12]   G. A. Flamme, M. R. Stephenson, and K. Deiters. "Typical noise exposure in daily life". In: *US National Library of Medicine* (Feb. 2012). URL: https://www.ncbi.nlm.nih.gov/pmc/articles/PMC4685462/ (cit. on p. 7).

[Gar+97]   S. D. Garvey, J. E. Penny, M. J. Friswell, and C. N. Glew. "Modelling the vibrational behaviour of stator cores of electrical machines with a view to successfully predicting machine noise". In: *IEE Colloquium on Modeling the Performance of Electrical Machines (Digest No: 1997/166)*. Apr. 1997, pp. 3/1–313. DOI: 10.1049/ic:19970897. URL: http://ieeexplore.ieee.org/document/643888/ (cit. on pp. 4, 94, 158).

[Ger+01]   F. Gerard, M. Tournour, N. E. Masri, L. Cremersa, M. Felice, and A. Selmane. "Acoustic Transfer Vectors for Numerical Modeling of Engine Noise". In: *Society of Automotive Engineers* (2001) (cit. on p. 165).

[GWC06]   J. F. Gieras, C. Wang, and J. Chao Lai. *Noise of Polyphase Electric Motors*. CRC Press of Taylor & Francis Group, LLC, 2006 (cit. on pp. 4, 8, 29, 31, 32, 34, 38, 41, 46, 48, 49, 51, 61, 91, 96, 163, 164).

[GMA99]   J. Gojko, D. Momir, and O. Aleksandar. "Skew and linear rise of MMF across slot modelling-winding function approach". In: *Energy Conversion, IEEE Transactions on* 14.3 (Sept. 1999), pp. 315–320. ISSN: 0885-8969. DOI: 10.1109/60.790876 (cit. on pp. 34, 52).

[Hau08]   F. Haugen. *Derivation of a Discrete-Time Lowpass Filter*. Mar. 2008. URL: http://techteach.no/simview/lowpass_filter/doc/filter_algorithm.pdf (cit. on p. 15).

[HGP12]   M. Hausmann, N. Grass, and B. Piepenbreier. "Harmonic compensation in a load emulation system using different control techniques". In: *Intelec 2012*. Sept. 2012, pp. 1–7. DOI: 10.1109/INTLEC.2012.6374496 (cit. on pp. 128, 130).

[HRC05]   H. Henao, H. Razik, and G.-A. Capolino. "Analytical approach of the stator current frequency harmonics computation for detection of induction machine rotor faults". In: *Industry Applications, IEEE Transactions on* 41.3 (May 2005), pp. 801–807. ISSN: 0093-9994. DOI: 10.1109/TIA.2005.847320 (cit. on pp. 32, 34).

[HW17]   T. Hu and C. Wang. "Frequency components analysis of an induction machine under different fault conditions". In: *2017 IEEE Power Energy Society General Meeting*. July 2017, pp. 1–5. DOI: 10.1109/PESGM.2017.8274726 (cit. on p. 125).

[Inc17a]   A. Inc. *ANSYS - Structural Analysis*. 2017. URL: www.ansys.com (cit. on p. 162).

[Inc17b]   M. Inc. *MathWorks - Accelerating the pace of engineering and science*. 2017. URL: https://de.mathworks.com/products.html?s_tid=gn_ps (cit. on p. 23).

[Inf05]   Infineon. "Optimized Space Vector Modulation and Over-modulation with the XC866". In: *Application Note* (2005). URL: http://www.infineon.com/dgdl/ap0803620_Space_Vector_Modulation.pdf?fileId=db3a304412%20b407950112b4199b7b28b3 (cit. on p. 19).

[JST16]   M. Jecmenica, D. Sosic, and M. Terzic. "Estimation of deep-bar induction motor rotor parameters using heuristic methods of optimization". In: *Mediterranean Conference on Power Generation, Transmission, Distribution and Energy Conversion (MedPower 2016)*. Nov. 2016, pp. 1–8. DOI: 10.1049/cp.2016.1111 (cit. on p. 23).

[JDP01]   G. Joksimovic, M. Djurovic, and J. Penman. "Cage Rotor MMF: Winding Function Approach". In: *Power Engineering Review, IEEE* 21.4 (Apr. 2001), pp. 64–66. ISSN: 0272-1724. DOI: 10.1109/MPER.2001.4311316 (cit. on p. 34).

[JSO17]   C. JSOL. *JMAG - Simulation Technology for Electromechnical Design*. 2017. URL: https://www.jmag-international.com/ (cit. on p. 23).

[Kaw+09]   Y. Kawase, T. Yamaguchi, Z. Tu, N. Toida, N. Minoshima, and K. Hashimoto. "Effects of Skew Angle of Rotor in Squirrel-Cage Induction Motor on Torque and Loss Characteristics". In: *IEEE Transactions on Magnetics* 45.3 (Mar. 2009), pp. 1700–1703. ISSN: 0018-9464. DOI: 10.1109/TMAG.2009.2012785 (cit. on p. 48).

[Ken12]   R. Kennel. *Raumzeigermodulation*. 2012. URL: https://www.eal.ei.tum.de/ fileadmin/tueieal/www/courses/UEEML/tutorial/RZM.pdf (cit. on p. 19).

[Ken13]   R. Kennel. *Power Electronics - Exercise: Space Vector Modulation*. 2013. URL: https://www.eal.ei.tum.de/fileadmin/tueieal/www/courses/PE/ tutorial/2013-2014-W/08_Space_vector_modulation.pdf (cit. on p. 19).

[Ken17]   R. Kennel. *Leistungselektronik, Grundlagen und Stardaardanwendungen*. Jan. 2017 (cit. on p. 18).

[Klo98]   G. Kloos. "Magnetostatic Maxwell stresses and magnetostriction". English. In: *Electrical Engineering* 81.2 (1998), pp. 77–80. ISSN: 0948-7921. DOI: 10.1007/ BF01237889. URL: http://dx.doi.org/10.1007/BF01237889 (cit. on p. 23).

[Kön07]   W. König. "Wie laut sind unsere Autos wirklich?" In: *Auto Bild* (Aug. 2007). URL: http://www.autobild.de/bilder/auto-bild-laermtest-376890. html#bild3 (cit. on p. 8).

[Kot+17a]   P. Kotter, W. Bischof, R. Kennel, O. Zirn, and K. Wegener. "Noise-vibration-harshness-modeling and analysis of induction drives in E-mobility applications". In: *2017 IEEE International Electric Machines and Drives Conference (IEMDC)*. May 2017, pp. 1–8. DOI: 10.1109/IEMDC.2017.8002148 (cit. on pp. 92, 95–98, 107, 111).

[Kot+16]   P. Kotter, B. Callan-Bartkiw, M. Boesing, K. Wegener, J. Berkemer, and O. Zirn. "Noise-vibration-harshness simulation of ultralight vehicle traction drives based on a universal modelling approach". In: *8th IET International Conference on Power Electronics, Machines and Drives (PEMD 2016)*. Apr. 2016, pp. 1–6. DOI: 10.1049/cp.2016.0218 (cit. on pp. 95, 107).

[KZW16]   P. Kotter, O. Zirn, and K. Wegener. "Efficient Noise-Vibration-Harshness Modelling of Servo- and Traction Drives". In: 49 (Dec. 2016), pp. 330–338 (cit. on pp. 95, 96, 98).

[Kot+17b]   P. Kotter, W. M. Bischof, C. Köpf, and K. Wegener. "Efficient noise-vibration-harshness-modeling for squirrel-cage induction drives in EV applications". In: *17. Internationales Stuttgarter Symposium*. 2017 (cit. on pp. 5, 92, 95, 96, 98, 107, 111, 112).

[LRO94]   T. I. Laakso, J. Ranta, and S. J. Ovaska. "Design and implementation of efficient IIR notch filters with quantization error feedback". In: *IEEE Transactions on Instrumentation and Measurement* 43.3 (June 1994), pp. 449–456. ISSN: 0018-9456. DOI: 10.1109/19.293466. URL: http://ieeexplore.ieee.org/document/293466/ (cit. on p. 16).

[Lac05]   R. Lach. "Magnetische Geräuschemission umrichtergespeister Käfigläufer - Asynchronmaschinen". PhD thesis. Universität Dortmund, 2005 (cit. on pp. 29, 32).

[Lam18]   F. Lambert. *Tesla motor designer explains Model 3s transition to permanent magnet motor*. Feb. 2018. URL: https://electrek.co/2018/02/27/tesla-model-3-motor-designer-permanent-magnet-motor/ (cit. on p. 3).

[LBD14]   T. Lange, M. Boesing. and R. W. De Doncker. "Measurement-Parameterized Synchronous Machine Model with Spatial-Harmonics". In: *Proceedings of the 17th International Conference on Electrical Machines and Systems (ICEMS)*. Zhejiang University. Hangzhou, China, Oct. 2014 (cit. on p. 54).

[Las+07]   C. Lascu, L. Asiminoaei, I. Boldea, and F. Blaabjerg. "High Performance Current Controller for Selective Harmonic Compensation in Active Power Filters". In: *Power Electronics, IEEE Transactions on* 22.5 (Sept. 2007), pp. 1826–1835. ISSN: 0885-8993. DOI: 10.1109/TPEL.2007.904060 (cit. on p. 130).

[Le +09]   J. Le Besnerais, V. Lanfranchi, M. Hecquet, G. Friedrich, and P. Brochet. "Characterisation of radial vibration force and vibration behaviour of a pulse-width modulation-fed fractional-slot induction machine". In: *Electric Power Applications, IET* 3.3 (May 2009), pp. 197–208. ISSN: 1751-8660. DOI: 10.1049/iet-epa.2008.0099 (cit. on pp. 31, 38, 96).

[Lee61]   C. Lee. "Saturation Harmonics of Polyphase Induction Machines". In: *Power Apparatus and Systems, Part III. Transactions of the American Institute of Electrical Engineers* 80.3 (Apr. 1961), pp. 597–603. ISSN: 0097-2460. DOI: 10.1109/AIEEPAS.1961.4501100 (cit. on p. 41).

[LT74]   M. Lees and C. Tindall. "Field-theory analysis of saturation harmonics in induction machines". In: *Electrical Engineers, Proceedings of the Institution of* 121.4 (Apr. 1974), pp. 276–280. ISSN: 0020-3270. DOI: 10.1049/piee.1974.0053 (cit. on p. 41).

[Lim+09]   L. Limongi, R. Bojoi, G. Griva, and A. Tenconi. "Digital current-control schemes". In: *Industrial Electronics Magazine, IEEE* 3.1 (Mar. 2009), pp. 20–31. ISSN: 1932-4529. DOI: 10.1109/MIE.2009.931894 (cit. on p. 126).

[LTB06]   M. Liserre, R. Teodorescu, and F. Blaabjerg. "Multiple harmonics control for three-phase grid converter systems with the use of PI-RES current controller in a rotating frame". In: *Power Electronics, IEEE Transactions on* 21.3 (May 2006), pp. 836–841. ISSN: 0885-8993. DOI: 10.1109/TPEL.2006.875566 (cit. on p. 130).

[Liw42]   M. M. Liwschitz. "Field Harmonics in Induction Motors". In: *American Institute of Electrical Engineers, Transactions of the* 61.11 (Nov. 1942), pp. 797–803. ISSN: 0096-3860. DOI: 10.1109/T-AIEE.1942.5058444 (cit. on p. 52).

[Mal00]   K. C. Maliti. "Modelling and Analysis of Magnetic Noise in Squirrel-Cage Induction Motors". PhD thesis. KTH Högskoletryckeriet Stockholm, Royal Institute of Technology, 2000 (cit. on pp. 31, 34, 38, 41, 42, 52).

[Mas14]   R. Massey. "Silent but deadly: EU rules all electric cars must make artificial engine noise". In: *Daily Mail* (2014). URL: http://www.dailymail.co.uk/news/article-2595451/Silent-deadly-EU-rules-electric-cars-make-artificial-engine-noise-fears-kill-unsuspecting-pedestrians.html (cit. on p. 7).

[MTG01]   C. McCulloch, M. Tournour, and P. Guisset. "Modal Acoustic Transfer Vectors Make Acoustic Radiation Models Practical for Engines and Rotating Machinery". In: *LMS International* (2001) (cit. on p. 165).

[McK11]   McKinsey. "Prognose zum weltweiten Umsatz mit Elektrofahrzeugen von 2010 bis 2030". In: *Handelsblatt* (Feb. 2011), p. 27. URL: https://de.statista.com/statistik/daten/studie/173157/umfrage/prognose-zum-weltweiten-umsatz-mit-elektrofahrzeugen/ (cit. on p. 2).

[MN83]   J. Melkebeek and D. Novotny. "The Influence of Saturation on Induction Machine Drive Dynamics". In: *Industry Applications, IEEE Transactions on* IA-19.5 (Sept. 1983), pp. 671–681. ISSN: 0093-9994. DOI: 10.1109/TIA.1983.4504275 (cit. on pp. 41, 52).

[MUR06]   Mohan, Undeland, and Riobbins. *Power Electronics*. Wiley India Pvt. Ltd, 2006 (cit. on p. 19).

[ML92]   J. Moreira and T. Lipo. "Modeling of saturated AC machines including air gap flux harmonic components". In: *Industry Applications, IEEE Transactions on* 28.2 (Mar. 1992), pp. 343–349. ISSN: 0093-9994. DOI: 10.1109/28.126740 (cit. on pp. 41, 42, 52).

[MK96]   D. Morgan and S. Kuo. *Active noise control systems. Algorithms and DSP implementations*. Wiley New York, 1996 (cit. on p. 127).

[MM04]   G. Müller and M. Möser. *Taschenbuch der Technischen Akustik*. Springer-Verlag Berlin Heidelberg, 2004 (cit. on p. 9).

[MP06]   G. Müller and B. Ponick. *Grundlagen elektrischer Maschinen*. WILEY-VCH Verlag GmbH & Co. KGaA, Weinheim, 2006 (cit. on pp. 3, 9, 10).

[MP09]   G. Müller and B. Ponick. *Theorie elektrischer Maschinen*. WILEY-VCH Verlag GmbH & Co. KGaA, Weinheim, 2009 (cit. on pp. 23, 48, 49, 51, 61).

[MVP08]   G. Müller, K. Vogt, and B. Ponick. *Berechnung elektrischer Maschinen*. 18th ed. WILEY-VCH Verlag GmbH & Co. KGaA, Weinheim, 2008 (cit. on pp. 3, 23–26, 28, 29, 31, 110, 111, 116, 125).

[NBT02]   S. Nandi, R. M. Bharadwaj, and H. A. Toliyat. "Mixed eccentricity in three phase induction machines: analysis, simulation and experiments". In: *Conference Record of the 2002 IEEE Industry Applications Conference. 37th IAS Annual Meeting (Cat. No.02CH37344)*. Vol. 3. Oct. 2002, 1525–1532 vol.3. DOI: 10.1109/IAS.2002.1043737 (cit. on p. 125).

[Nan+11]   S. Nandi, T. C. Ilamparithi, S. B. Lee, and D. Hyun. "Detection of Eccentricity Faults in Induction Machines Based on Nameplate Parameters". In: *IEEE Transactions on Industrial Electronics* 58.5 (May 2011), pp. 1673–1683. ISSN: 0278-0046. DOI: 10.1109/TIE.2010.2055772 (cit. on p. 125).

[NZE05]   P. Ngatchou, A. Zarei, and A. El-Sharkawi. "Pareto Multi Objective Optimization". In: *Proceedings of the 13th International Conference on, Intelligent Systems Application to Power Systems*. Nov. 2005, pp. 84–91. DOI: 10.1109/ISAP.2005.1599245 (cit. on p. 2).

[Nyq28]   H. Nyquist. "Certain Topics in Telegraph Transmission Theory". In: *Transactions of the American Institute of Electrical Engineers* 47.2 (Apr. 1928), pp. 617–644. ISSN: 0096-3860. DOI: 10.1109/T-AIEE.1928.5055024 (cit. on p. 10).

[Ope06]   OpenCourseWare. "Discrete-Time Signal Processing". In: *Department of Electrical Engineering and Computer Science* (2006). URL: https://ocw.mit.edu/courses/electrical-engineering-and-computer-science/6-341-discrete-time-signal-processing-fall-2005/lecture-notes/lec08.pdf (cit. on p. 15).

[OM03]   K. Oskar and S. Matthias. *Algorithmen, Schnittstellen und Werkzeuge zur Audiobearbeitung*. TU Muenchen, Fakultaet fuer Informatik, 2003 (cit. on p. 127).

[PLF12a]   P. Pellerey, V. Lanfranchi, and G. Friedrich. "Coupled Numerical Simulation Between Electromagnetic and Structural Models. Influence of the Supply Harmonics for Synchronous Machine Vibrations". In: *Magnetics, IEEE Transactions on* 48.2 (Feb. 2012), pp. 983–986. ISSN: 0018-9464. DOI: 10.1109/TMAG.2011.2175714 (cit. on p. 23).

[PLF12b]   P. Pellerey, V. Lanfranchi, and G. Friedrich. "Numerical simulations of rotor dynamic eccentricity effects on synchronous machine vibrations for full run up". In: *Electrical Machines (ICEM), 2012 XXth International Conference on*. Sept. 2012, pp. 3008–3014. DOI: 10.1109/ICElMach.2012.6350316 (cit. on p. 46).

[PEE15]   A. Polat, Y. D. Ertugrul, and L. T. Ergene. "Static, dynamic and mixed eccentricity of induction motor". In: *2015 IEEE 10th International Symposium on Diagnostics for Electrical Machines, Power Electronics and Drives (SDEMPED)*. Sept. 2015, pp. 284–288. DOI: 10.1109/DEMPED.2015.7303703 (cit. on p. 46).

[QD15]   N. P. Quang and J. Dittrich. *Inverter Control with Space Vector Modulation*. Springer-Verlag Berlin Heidelberg, 2015 (cit. on p. 19).

[Ram+16]   K. Ramakrishnan, S. Stipetic, M. Gobbi, and G. Mastinu. "Multi-objective optimization of electric vehicle powertrain using scalable saturated motor model". In: *2016 Eleventh International Conference on Ecological Vehicles and Renewable Energies (EVER)*. Apr. 2016, pp. 1–6. DOI: `10.1109/EVER.2016.7476430` (cit. on p. 2).

[RCO13a]   G. A. Ramos, R. Costa-Castello, and J. M. Olm. *Digital Repetitive Control under Varying Frequency Conditions*. Springer Verlag Berlin Heidelberg, 2013 (cit. on p. 126).

[RCO13b]   G. A. Ramos, R. Costa-Castello, and J. M. Olm. *Digital Repetitive Control under Varying Frequency Conditions*. Springer-Verlag Berlin Heidelberg, 2013 (cit. on p. 126).

[Ria]   M. Riaz. *SQUIRREL-CAGE INDUCTION MACHINE*. `http://www.ece.umn.edu/users/riaz/animations/sqmoviemotgen.html` (cit. on p. 31).

[Rip07]   W. Rippel. *Induction Versus DC Brushless Motors*. Jan. 2007. URL: `https://www.tesla.com/de_DE/blog/induction-versus-dc-brushless-motors` (cit. on p. 3).

[Roi09]   J. Roivainen. "UNIT-WAVE RESPONSE-BASED MODELING OF ELECTROMECHANICAL NOISE AND VIBRATION OF ELECTRICAL MACHINES". PhD thesis. Helsinki University of Technology, 2009 (cit. on pp. 97, 162).

[RJK94]   S. C. D. Roy, S. B. Jain, and B. Kumar. "Design of digital FIR notch filters". In: *IEE Proceedings - Vision, Image and Signal Processing* 141.5 (Oct. 1994), pp. 334–338. ISSN: 1350-245X. DOI: `10.1049/ip-vis:19941057`. URL: `http://ieeexplore.ieee.org/document/331652/` (cit. on p. 16).

[SA12]   B. M. Saied and A. J. Ali. "Determination of deep bar cage rotor induction machine parameters based on finite element approach". In: *2012 First National Conference for Engineering Sciences (FNCES 2012)*. Nov. 2012, pp. 1–6. DOI: `10.1109/NCES.2012.6740481` (cit. on p. 23).

[SH11]   S. Sarkka and A. Huovilainen. "Accurate Discretization of Analog Audio Filters With Application to Parametric Equalizer Design". In: *IEEE Transactions on Audio, Speech, and Language Processing* 19.8 (Nov. 2011), pp. 2486–2493. ISSN: 1558-7916. DOI: `10.1109/TASL.2011.2144970` (cit. on p. 16).

[Sch12]   D. Schröder. *Leistungselektronische Schaltungen*. Ed. by 3. Springer-Verlag Berlin Heidelberg, 2012 (cit. on p. 18).

[Sch15]   D. Schröder. *Elektrische Antriebe - Regelung von Antriebssystemen*. Ed. by 4. Springer-Verlag Berlin Heidelberg, 2015 (cit. on pp. 17, 18).

[Sch10]   D. Schroeder. *Intelligente Verfahren*. Springer-Verlag Berlin Heidelberg, 2010 (cit. on p. 127).

[Sch14]   C. M. Schwarzer. "So sauber ist das Elektroauto". In: *Zeit Online* (Jan. 2014). URL: http://www.zeit.de/mobilitaet/2014-01/elektroauto-energiebilanz (cit. on p. 2).

[SEM04]   D. E. Seborg, T. F. Edgar, and D. A. Mellinchamp. *Process, Dynamics and Control.* John Wiley & Sons, 2004. URL: https://www.google.de/url? sa=t&rct=j&q=&esrc=s&source=web&cd=9&ved=0ahUKEwjFoZvo- ufTAhWE1hoKHdQSAZQCFghbMAg&url=http%3A%2F%2Fbank.engzenon.com% 2Fdownload%2F53050b2d-9aa0-4a58-af96-58f4c0e8c6f8%2FProcess_ Dynamics_and_Control_Seborg_2nd_edition.pdf&usg=AFQjCNGRHYWsWJr- 17oYLyEOI5RdY3Azwg&cad=rja (cit. on p. 18).

[Sei92]   H. O. Seinsch. *Oberfelderscheinungen in Drehfeldmaschinen.* B. G. Teubner Stuttgart, 1992 (cit. on pp. 29, 42, 49, 51, 61, 92).

[Sha49]   C. E. Shannon. "Communication in the Presence of Noise". In: *Proceedings of the IRE* 37.1 (Jan. 1949), pp. 10–21. ISSN: 0096-8390. DOI: 10.1109/JRPROC. 1949.232969. URL: http://ieeexplore.ieee.org/document/1697831/ (cit. on pp. 15, 17).

[Sop05]   J. O. Sophocles. "High-Order Digital Parametric Equalizer Design". In: *Journal of the Audio Engineering Society* 53 (2005), pp. 1026–1046. URL: http://www. ece.rutgers.edu/~orfanidi/ece346/hpeq.pdf (cit. on p. 15).

[Sta10]   C. .-.-. E. C. F. Standardization. "Determination of sound power levels and sound energy levels of noise sources using sound pressure - Engineering methods for an essentially free field over a reflecting plane". In: *European Standard EN ISO 3744:2010* (2010) (cit. on p. 163).

[Sta]   T. Stathaki. *LTI Discrete-Tiime Systems in Transform Domain.* URL: http: //www.commsp.ee.ic.ac.uk/~tania/teaching/dsp/Lecture%207%20LTI% 20Discrete-Time%20Systems%20in%20the%20Transform%20Domain.pdf (cit. on p. 15).

[SBS01]   C. D. R. Suhash, K. Balbir, and B. J. Shail. "FIR NOTCH FILTER DESIGN - A REVIEW". In: *Facta Universitatis series: Electronics and Energetics* 14 (2001), pp. 295–327 (cit. on p. 16).

[ST77]   M. N. S. Swamy and K. S. Thyagarajan. "Frequency transformations for digital filters". In: *Proceedings of the IEEE* 65.1 (Jan. 1977), pp. 165–166. ISSN: 0018- 9219. DOI: 10.1109/PROC.1977.10441. URL: http://ieeexplore.ieee.org/ document/1454710/ (cit. on p. 16).

[Sys17]   D. Systems. *Simula - Abaqus Unified FEA.* 2017. URL: https://www.3ds.com/ de/produkte-und-services/simulia/produkte/abaqus/ (cit. on p. 162).

[Sys03]   G. I. Systems. "SVM Space Vector Modulation". In: *White Paper* (2003). URL: http://www.lpsolutions-inc.com/resources/SVMSpaceVectorModulation. pdf (cit. on p. 19).

[TA01]   A. Tenhunen and A. Arkkio. "Modelling of induction machines with skewed rotor slots". In: *IEE Proceedings - Electric Power Applications* 148.1 (Jan. 2001), pp. 45–50. ISSN: 1350-2352. DOI: 10.1049/ip-epa:20010036 (cit. on p. 48).

[TL94]   P. L. Timar and J. C. S. Lai. "Acoustic noise of electromagnetic origin in an ideal frequency-converter-driven induction motor". In: *IEE Proceedings - Electric Power Applications* 141.6 (Nov. 1994), pp. 341–346. ISSN: 1350-2352. DOI: 10.1049/ip-epa:19941342 (cit. on p. 164).

[vdGie11]   M. van der Giet. "Analysis of electromagnetic acoustic noise excitations". PhD thesis. RWTH Aachen, 2011. URL: http://www.shaker.de/Online-Gesamtkatalog-Download/2017.04.20-17.38.10-194.39.218.10-rad87590.tmp/3-8322-9973-4_INH.PDF (cit. on pp. 4, 97, 162, 164).

[VBG92]   D. Verdyck, R. Belmans, and W. Geysen. "An acoustic model for a permanent magnet machine: modal shapes and magnetic forces". In: *Conference Record of the 1992 IEEE Industry Applications Society Annual Meeting*. Oct. 1992, 292–299 vol.1. DOI: 10.1109/IAS.1992.244281. URL: http://ieeexplore.ieee.org/document/244281/ (cit. on pp. 4, 92).

[VBG93]   D. Verdyck, R. Belmans, and W. Geysen. "An approach to modelling of magnetically excited forces in electrical machines". In: *IEEE Transactions on Magnetics* 29.2 (Mar. 1993), pp. 2032–2035. ISSN: 0018-9464. DOI: 10.1109/20.250809 (cit. on p. 92).

[VB94]   S. P. Verma and A. Balan. "Determination of radial-forces in relation to noise and vibration problems of squirrel-cage induction motors". In: *Energy Conversion, IEEE Transactions on* 9.2 (June 1994), pp. 404–412. ISSN: 0885-8969. DOI: 10.1109/60.300130 (cit. on p. 61).

[vPSH16]   G. von Pfingsten, S. Steentjes, and K. Hameyer. "Transient approach to model operating point dependent losses in saturated induction machines". In: *2016 XXII International Conference on Electrical Machines (ICEM)*. Sept. 2016, pp. 626–632. DOI: 10.1109/ICELMACH.2016.7732591 (cit. on p. 116).

[WS85]   B. Widrow and S. D. Stearns. *Adaptive signal processing*. Prentice-Hall signal processing series. Englewood Cliffs, N.J. : Prentice-Hall, c1985., 1985. ISBN: 0130040290. URL: http://search.ebscohost.com/login.aspx?direct=true&db=edshlc&AN=edshlc.000557910-4&site=eds-live (cit. on p. 127).

[Wil71]   J. Willems. "Space harmonics in unified electrical-machine theory". In: *Electrical Engineers, Proceedings of the Institution of* 118.10 (Oct. 1971), pp. 1408–1412. ISSN: 0020-3270. DOI: 10.1049/piee.1971.0264 (cit. on p. 61).

[WT06]   A. Williams and F. Taylor. *Electronic Filter Design Handbook*. MacGraw-Hill Handbooks, 2006. URL: http://kvetakov.net/~mira/literatura/Electronic%20Filter%20Design%20Handbook%204th%20Ed.pdf (cit. on p. 15).

[Wol+15] C. Wolz, M. Greger, B. Wüchner, J. Kempkes, and U. Schäfer. "Calculation of nonlinear flux linkage and torque for permanent magnet DC motors". In: *2015 IEEE International Electric Machines Drives Conference (IEMDC)*. May 2015, pp. 147–153. DOI: 10.1109/IEMDC.2015.7409052 (cit. on p. 49).

[Yep+10] A. G. Yepes, F. D. Freijedo, J. D. Gandoy, O. Lopez, and J. Malvar. "Effects of Discretization Methods on the Performance of Resonant Controllers". In: *IEEE Transactions on Power Electronics* 25.7 (July 2010), pp. 1692–1712. ISSN: 0885-8993. DOI: 10.1109/TPEL.2010.2041256. URL: http://ieeexplore.ieee.org/stamp/stamp.jsp?arnumber=5398914 (cit. on pp. 130, 132).

[Yep+11] A. Yepes, F. Freijedo, O. Lopez, and J. Doval-Gandoy. "High-Performance Digital Resonant Controllers Implemented With Two Integrators". In: *Power Electronics, IEEE Transactions on* 26.2 (Feb. 2011), pp. 563–576. ISSN: 0885-8993. DOI: 10.1109/TPEL.2010.2066290 (cit. on p. 130).

[Yep11] A. G. Yepes. "Digital Resonant Current Controllers For Voltage Source Converters". PhD thesis. 2011 (cit. on p. 130).

[Yi+13] H. Yi, F. Zhuo, Y. Li, Y. Zhang, and W. Zhan. "Comparison Analysis of Resonant Controllers for Current Regulation of Selective Active Power Filter with Mixed Current Reference". In: *Journal of Power Electronics* 13.5 (2013), pp. 861–876 (cit. on p. 130).

[ZW02] K. Zhou and D. Wang. "Relationship between space-vector modulation and three-phase carrier-based PWM: a comprehensive analysis [three-phase inverters]". In: *IEEE Transactions on Industrial Electronics* 49.1 (Feb. 2002), pp. 186–196. ISSN: 0278-0046. DOI: 10.1109/41.982262 (cit. on p. 19).

[ZHB99] D. Zmood, D. Holmes, and G. Bode. "Frequency domain analysis of three phase linear current regulators". In: *Industry Applications Conference, 1999. Thirty-Fourth IAS Annual Meeting. Conference Record of the 1999 IEEE*. Vol. 2. 1999, 818–825 vol.2. DOI: 10.1109/IAS.1999.801601 (cit. on p. 130).

[Zöl97] U. Zölzer. *Digital Audio Signal Processing*. John Wiley & Sons, LTD, 1997. URL: http://ultra.sdk.free.fr/docs/DxO/[DSP%20-%20audio]%20-%20Digital%20Audio%20Signal%20Processing%20-%20Zolzer.pdf (cit. on p. 16).

[ZSW16] ZSW. "Weltweite Bestandsentwicklung von Elektroautos in den Jahren 2012 bis 2016". In: *Zentrum für Sonnenenergie- und Wasserstoff-Forschung Baden-Württemberg (ZSW)* 05/2016 (May 2016). URL: https://de.statista.com/statistik/daten/studie/168350/umfrage/bestandsentwicklung-von-elektrofahrzeugen/ (cit. on p. 2).